ACS SYMPOSIUM SERIES 787

Fluorinated Surfaces, Coatings, and Films

David G. Castner, EDITOR

University of Washington

David W. Grainger, EDITOR

Colorado State University

American Chemical Society, Washington, DC

Library of Congress Cataloging-in-Publication Data

Fluorinated surfaces, coatings, and films / David G. Castner, editor, David W. Grainger, editor.

    p. cm.—(ACS symposium series ; 787)

    Includes bibliographical references and index.

    ISBN 0–8412–3623–2

    1. Fluoropolymers—Congresses. 2. Coatings—Congresses. 3. Thin films—Congresses.

    I. Castner, David Gordon, 1952- II. Grainger, David W., 1961- III. Series.

QD383 .F8 F55   2001
667′.9—dc21
                                            00–53114

The paper used in this publication meets the minimum requirements of American National Standard for Information Sciences—Permanence of Paper for Printed Library Materials, ANSI Z39.48–1984.

Foreword

The ACS Symposium Series was first published in 1974 to provide a mechanism for publishing symposia quickly in book form. The purpose of the series is to publish timely, comprehensive books developed from ACS sponsored symposia based on current scientific research. Occasionally, books are developed from symposia sponsored by other organizations when the topic is of keen interest to the chemistry audience.

Before agreeing to publish a book, the proposed table of contents is reviewed for appropriate and comprehensive coverage and for interest to the audience. Some papers may be excluded to better focus the book; others may be added to provide comprehensiveness. When appropriate, overview or introductory chapters are added. Drafts of chapters are peer-reviewed prior to final acceptance or rejection, and manuscripts are prepared in camera-ready format.

As a rule, only original research papers and original review papers are included in the volumes. Verbatim reproductions of previously published papers are not accepted.

ACS Books Department

Contents

vi

Preface

The technological value of placing low energy perfluoroalkyl groups at surfaces has been recognized since the pioneering studies of perfluorinated surfactants by Zisman and colleagues at the U.S. Naval Research Lab just after World War II. Plunkett's earlier serendipitous discovery of polytetrafluoroethylene (PTFE, DuPont's Teflon™) provided an intellectual stimulus and subsequent commercial source for interesting, completely fluorinated polymeric material. Its expense, poor processing properties, and mechanical deficiencies limited applications for PTFE, however. The general interest in perfluorinated materials chemistry as a surface energy reducing agent, lubricant, optical cladding, dielectric barrier, or selectively permeable membrane in thin-film form prompted active research and development efforts in two areas of perfluorinated film chemistry: lower cost fluoropolymers with improved film processing and small perfluoroalkyl-containing molecules capable of forming coherent, stable films and coatings. Both technical approaches recognized that perfluorinated chemistry is inherently expensive and should therefore be placed only where it is required: at the interface. These efforts have now yielded a broad array of commercialized perfluorinated coating products with versatile properties and applications both for complex technologies (communications, aerospace, military) and consumer markets (e.g., Scotchgard™ and Stainmaster™ treatments, Gore-tex™ fabrics, and coated cookware).

Although synthetic chemistry surrounding perfluoroalkyl materials continues to be a focus for both academic and industrial scientists, applications of perfluorinated materials has also benefited enormously from advancements in surface analytical methods and instrumentation. For example, use of X-ray photoelectron spectroscopy (XPS or ESCA) is ideally suited for profiling the chemical composition of the outer 90Å of films and coatings. Yet, polychromatic X-rays used prior to development of the newer XPS monochromatic systems rapidly degraded halogenated surfaces, making accurate quantitation of fluorine in coatings almost impossible. Modern surface analytical methods such as XPS, static time-of-flight-secondary ion mass spectrometry (ToF-SIMS), infrared spectroscopies, scanning probe microscopy (SPM), and near edge X-ray absorption fine structure (NEXAFS) can elucidate both the chemical composition and structure of fluorinated surfaces in unprecedented detail. The complementary information from these methods provides spatial distribution of chemistry on surfaces, specific chemical speciation within the top few molecular layers at a surface, and some structural features associated with orientation of chains in the surface zone.

Because the valuable technological properties of fluorinated films and coatings are associated both with perfluoroalkyl chain chemistry and chain orientation, methods and techniques that control or place specific perfluoroalkyl groups at the interface are a current focus. Characterization methods that elucidate surface structure of such chemistry and relate them to interfacial properties (e.g., friction, lubrication, dielectric constant, and wetting) are also critical. In this regard, alternative fluorinated materials (e.g., perfluoroalkyl SF_5-functional terminal chemistry, or mixed hydrocarbon–perfluorocarbon molecules) and new surface orientation-sensitive methods (synchrotron-based NEXAFS) are being introduced to the field. Both should facilitate advancement of the understanding of structure–chemistry–property relationships in fluorinated coatings and films, as well as expand the menu of materials and techniques available to further develop, manipulate, or improve these materials in specific interfacial applications.

This book was published as a result of an American Chemical Society (ACS) Division of Polymer Chemistry, Inc. sponsored symposium entitled "Fluorinated Surfaces, Coatings, and Films" held at the ACS National meeting in Boston, MA in 1998 (see ACS POLY Preprints, 1998, 39(2)). This symposium was deliberately planned together with another ACS POLY Symposium entitled "Fluoropolymers" that was also held in Boston at the same time. Together, these symposia drew a substantial amount of international interest from scientists working in these important, related fluorinated materials areas. As editors, we have tried to capture the excitement, relevance, and scientific theme of that meeting within these pages. Beyond the introductory first chapter, technical contributions address a wide variety of technical topics necessary to understand the placement, analysis, and significance of perfluorinated chemistry at interfaces, sometimes within the context of an industrially relevant problem or application. Surface structure of fluorinated surfactants in solid films or at the air–water interface as elucidated by XPS, microscopy, and surface tension measurements is complemented by interfacial behaviors of several novel perfluoropolymer coatings. Atomically-resolved perfluorinated film structures produced by SPM studies of fluorinated monolayers are contrasted with bulklike optical properties of fluorine-containing claddings. A significant set of contributions describes new developments in plasma-deposited or sputtered fluorinated chemistries, a topic of significant academic and commercial importance.

We believe that this book should prove a valuable resource for the professional chemist, engineer, and applications specialist seeking current thinking, know-how, and methods regarding fabrication, characterization, and properties of perfluorinated surfaces and interfaces. Both basic science and more industrially oriented research topics are represented across these areas. The level of presentation is appropriate for the advanced student, the technical manager, and the scientist seeking rapid integration to the compelling issues of this field.

DAVID G. CASTNER
National ESCA and Surface Analysis Center for Biomedical Problems
Departments of Bioengineering and Chemical Engineering
University of Washington
Seattle, WA 98195–1750

DAVID W. GRAINGER
Department of Chemistry
Colorado State University
Fort Collins, CO 80523–1872

Chapter 1

Fluorinated Coatings and Films: Motivation and Significance

David W. Grainger[1] and Charlie W. Stewart[2]

[1]Department of Chemistry, Colorado State University,
Fort Collins, CO 80523
[2]C. W. Stewart & Associates, 4 Jobs Lane, Newark, DE 19711

Fluorinated surfaces are sought for a number of technologies requiring low surface energy, low adhesion, chemical robustness, among other desired properties. Bulk fluorinated materials (e.g., polymers) can be used, but their relative expense and difficult processing problems preclude widespread application. Small molecule perfluorinated precursors are popular film-forming materials but, without crosslinking or stabilization *in situ*, lack the durability of higher molecular weight materials. Methods to fabricate perfluorinated coatings and films to impart the desirable properties of perfluorocarbons where they are most needed, without the disadvantages of handling and applying bulk perfluorinated materials are described in this volume. This chapter attempts to introduce the reader to the issues and general strategies described in detail in the following chapters.

Historical Perspective

Interests in and applications for fluorinated materials have advanced significantly since Plunkett first serendipitously discovered a slippery white powder inside a canister of tetrafluoroethylene (TFE) gas in 1938 that has since come to be known world-wide as Teflon®.(*1,2*) Prior to 1937, only three pure perfluorocarbons (carbon tetrafluoride, hexafluoroethane, and tetrafluoroethylene) (*3*) had been reported and characterized fully; research in fluorine chemistry was focused primarily on fluoroarenes prior to this point. Aliphatic fluorine chemistry, principally the work of Swarts at that time (*4*), had already been ostracized from the domain of mainstream organic chemistry, an unfortunate circumstance that is still relevant and damaging to progress today.(*5*) Several historic events early in this century served to catalyze work developing perfluorocarbons, including General Motors workers' early search for new refrigerants among aliphatic fluorides (*6*), Plunkett's discovery of TFE polymerization (*1*), Simons' and Blocks' 1939 report that amalgamated copper and mercury(II) fluoride yield perfluorocarbons (7), Manhattan Project secret efforts to use perfluorocarbons from Simons' lab as uranium derivatives in atomic bomb developments during World War II (8), and post-World War II escalation of perfluorocarbon research in both the Western hemisphere and Russia resulting from Cold War strategic defense purposes, particularly in fluoropolymer chemistry. A substantial Russian legacy in this area remains virtually entrapped and untapped in untranslated scientific literature to this day. The net result is the rapid commercialization of fluorinated materials in both consumer and technological markets that we witness currently.

Perfluorocarbon Properties Derive from Molecular Structure and Bonding

Technologically desirable characteristics of perfluorinated materials, particularly as surface coatings, derive from molecular properties and resulting associative behavior typically found in C-F and C-C bonds in perfluorocarbons. Incorporation of fluorine into organic materials, using a variety of chemical means, imparts dramatic changes to the material's physical and chemical properties. The value of these properties now recognized, many of the common commercial fluorinated materials are based on iterations of fluorinated carbon species, both low and high molecular weight. Fluorine's highest electronegativity creates quite polar C-F bonds capable of unpredictable and irregular hydrogen bonding patterns. Polar, partially perfluorinated materials are considerably different than either hydrocarbon or completely perfluorocarbon analogs. Single fluorine bonds with carbon are the strongest known with carbon, some 25 kcal mol^{-1} stronger than C-Cl (8). Together with the poor leaving group ability intrinsic to fluorine, this makes alkyl fluorides 10^2-10^6 times less reactive than the corresponding alkylchlorides in certain solvolysis and displacement reactions (8). Fluorination also influences adjacent bond energies. For example, addition of F strengthens adjacent aliphatic bonds: the CF_3–CF_3 bond is 10 kcal mol^{-1} stronger than the CH_3–CH_3 bond (8). Notably high thermal and chemical stabilities observed for perfluorinated materials result from these molecular bond properties. Fluorine's high ionization potential and low polarizability support a tendency for relatively weak intermolecular forces, low interfacial energies and low refractive indices in fluorinated materials. Partially fluorinated materials exhibit properties (e.g., boiling points) consistent with significantly higher intermolecular interactions.

Fluorine's larger radius compared to hydrogen has been the subject of some debate; this is important because of the steric implications for fluorinated materials structure and properties compared to analogous hydrocarbons. Well-accepted Pauling van der Waals radius values (1.35Å for F and 1.2Å for H) are still often-cited, with the accompanying false conclusion that in fact these atoms are the same size. Newer crystallographic data (9) and modified Taft steric parameters (10), however, now indicate that *fluorine and oxygen are more closely isosteric*, and that fluorine and the hydroxyl group are even *more closely* isosteric, (11) with hydrogen's steric radius not even closely comparable (8). Using these data, structural differences between perfluorocarbons and hydrocarbons can be rationalized. For example, rotational barriers for various fluorine-substituted bonds are significantly higher than barriers in analogous hydrocarbon systems (8). Significantly, the $-CF_3$ group is significantly larger than the $-CH_3$ group. Analysis supports more accurate comparison of $-CF_3$ with the *isopropyl* group, $-CH(CH_3)_2$.(12) While differences in piezoelectric properties for related partially fluorinated commercial polymers, poly(vinylidene fluoride) (PVDF) and ethylene-trifluoroethylene copolymer (ETFE), are attributed to zig-zag chain conformations that distinguish different C-F dipole alignments along the chain (8), perfluorocarbons with only C-F bonds are thought therefore to be stiffened and assume helical orientations in chains where C-F dipoles are distributed axially around the helix. (13-21) PTFE, for example, as a model for essentially infinite molecular weight perfluorocarbon chains, is known to have a rich phase diagram of several distinct helical solid phases (21)

Perfluorinated Surfaces.

Fluorinated surfaces derive their character from these unique molecular properties associated with C-F bonding chemistry that impart specific, unique chemistry and physics at interfaces. Two basic properties seem to be sought in the development of perfluorinated surfaces: low solid-state surface free energy, or surface tension, and chemical resistance and durability. Perfluorocarbons might be presumed, because of their intrinsically low intermolecular forces, to exhibit low interfacial tensions since surface tension (λ) is defined as molecular force per unit length (mN/m) opposing creation of new surface area. In fact, perfluorocarbons as a class of materials exhibit the lowest λ values for all organic liquids (shown in Table 1) and spread spontaneously ('wetting') over nearly all solid surfaces. Partially fluorinated materials exhibit interesting, irregular trends in surface tension when compared to fully hydrogenated analogs, but always exhibit surface tensions higher than their perfluorinated analogs.(22)

The effects of fluorination on surface energies in solids are similar to fluorine's influence on liquid interfacial behaviors. Table 1 shows that perfluorinated polymer surfaces exhibit low interfacial free energies (as indicated by Zisman γ_c values) that correlate directly with their utility as low energy, low adhesion, low friction surfaces. Substituting either hydrogen or another halogen for fluorine along the backbone results in significant increases in γ_c values as seen for comparisons of PTFE with either high-density polyethylene (HDPE), PVDF and poly(chlorotrifluoroethylene) (PCTFE) surfaces. For example, poly(CH_2CHF) has a γ_c of 28 mN/m, approaching that for HDPE (Table 1). The obvious conclusion from these data is that fluorination imparts lower surface energies to solids. The corollary is that fluorinated species prefer to reside on a material's surface for energetic reasons if allowed to assume the most thermodynamically stable position in a solid or liquid. This has particular importance in mixtures of components that have sufficient mobility (i.e., diffusivity, relaxation energy) to permit mass movement, migration, or reorganization: surface-enriched films of perfluorinated components are found to reside at surfaces even if present as dopants or minority components in bulk. This surface 'blooming' effect has origin in the surface activity (low interfacial tension) of perfluorocarbon species, as well as miscibility problems with components including hydrocarbons and monomer resins in bulk that favor surface partitioning.

Table 1: Surface tensions for perfluorocarbons and analogous hydrocarbons (8).*

	γ (mN/m)	
Substance	**Perfluorocarbon**	**Hydrocarbon**
PTFE[1]	18.5	NA
PVDF	25	--
PCTFE	31	--
HDPE[2]	NA	31
n-pentane	9.4	15.2
n-hexane	11.4	17.9
n-octane	13.6	21.1
Decalin	17.6	29.9
benzene	22.6	28.5

1: γ_c values

2: $\gamma_{l/v}$ values

*see also *Physical Properties of Polymers Handbook*, Mark, J.E., ed., AIP Press, Woodbury, NY, 1996, p.669.

Lowest surface free energies are generally attributed to ambient exposure of $-CF_3$ groups: the lowest γ_c value known was measured for close-packed, organized monolayers comprising the nearly vertically aligned $CF_3(CF_2)_{10}COOH$ molecule, a surface of closely packed $-CF_3$ groups.(23-25) Substituting only one hydrogen for one fluorine in the terminal, exposed $-CF_3$ group increases the surface free energy to less impressive value of 15 mN/m.(23) The realization that the $-CF_3$ group could be the 'Holy Grail' of the low surface free energy contest has also brought recognition that what lies beneath it must not always be perfluorinated if the $-CF_3$ group surface presentation is optimal. Surface free energy is a coupled function of both chemistry and organization. Surface density of $-CF_3$ groups can be maximized and surface tension thereby optimally reduced by exploiting $-CF_3$ group orientation, organization, and presentation. In fact, several strategies have sought to place and then orient $-CF_3$ groups on surfaces using various means. Because the surface density and organization of these groups is necessary to achieve low surface tensions, the $-CF_3$ group alone is not enough to reduce solid surface free energies to useful levels. Alignment of chains terminating in this chemistry is required: surface tension reduction proceeds to a minimum as R_f chain length approaches 8-10.(26) Coupling this with surface chain organization can often change this requirement: organized monolayers comprising $CF_3(CF_2)_n(CH_2)_{16}COOH$ reduce surface free energy to that observed for oriented monolayers of $CF_3(CF_2)_nCOOH$ when $n\exists6$.(*23-25*) Additionally, significant amounts of work have focused on orienting polymer films comprising perfluoroalkyl acrylates and methacrylates to enrich surfaces with the side chain terminal $-CF_3$ group.(*13,27-29*). Perfluoroalkyl-grafted polysiloxanes (*30*) also can produce side chain orientation to reduce surface tension via presentation of perfluorinated chemistry, sometimes with spontaneous perfluorinated side chain organization.(*31-36*)

This leads to the conclusion that desirable low surface tension and chemical inertness associated with fluorinated and perfluorinated species might be imparted to surfaces, without requiring the entire bulk material to be fluorinated. Perfluorinated chemistry is well-known to migrate to air-exposed interfaces spontaneously if permitted. Film formation and coatings from perfluorinated components form readily both as applied overlayers as well as from bulk mixtures due to this property. This has significant cost advantages since fluorination costs and associated processing are expensive. Additionally, often coatings properties are preferred as bulk fluorinated materials properties are less than ideal (e.g., PTFE). Emphasis on fluorinated coatings and films has therefore increased as technological drivers to provide impetus for new, improved and less expensive methods to put this chemistry on surfaces.(*37-39*)

Perfluorinated Polymers. While fluorinated coating materials comprise a wide variety of chemistries and physical embodiments, including surfactants, monomers, precursor chemicals and plasma depositions, fluorinated polymers are arguably the most well-known perfluorinated material coating.(*40*) As rapidly as new chemistries and properties of fluorinated polymers were reported, new potential applications were described. Table 2 summarizes some common abbreviations, base chemistry and suppliers for commercial fluoropolymers developed during the late 20th century. DuPont's tradename, Teflon®, actually represents a commercial family of several distinct fluoropolymer chemistries in addition to the original poly(tetrafluoroethylene) (PTFE). Other commercial suppliers market many other lines of fluorinated materials, primarily polymers, for both bulk and coating applications. Poly(tetrafluoroethylene is the largest volume commercial fluoropolymer. It is characterized by high molecular weight chains (10^6-10^7), high crystallinity, low solubility, high chemical inertness, high hydrophobicity, high creep or flow under stress, poor mechanical and machining properties, high opacity, poor processability, sensitivity to radiation and high bulk cost. Low polymer chain-chain cohesive energy imparts the slippery nature and surface lubricity characteristic of PTFE. Clearly there exist trade-offs in selecting such a material for a surface coating.

Because of traditionally poor mechanical and bulk performance of PTFE in addition to its bulk cost, demands for fluorinated materials have generally come from specific surface properties intrinsic to fluorinated surfaces: chemical robustness, tribology and frictional properties, surface non-wetting, and in some cases, optical properties (e.g., optical fiber cladding (*41,42*)). Chemical improvements to PTFE to achieve greater processing capabilities have diversified the markets for this material type. These strategies have introduced various copolymerizations with aliphatic monomers, deliberately induced branching, altered side chain chemistries, vinyl ether copolymers, and backbone alterations (e.g., vinylidene or perfluoroether copolymers) in attempting to reduce polymer bulk crystallinity while maintaining desirable PTFE properties. Many PTFE variants have resulted. Economic concerns regarding materials costs, however, have directed efforts to then achieve these desired interfacial properties using as little fluorinated component as possible. One option for polymers is to fluorinate their surfaces post-polymerization. Polyethylene (*43*), polysulfone (*44*), polystyrene (*44*), and cellulose (*45*) have all been surface-treated to introduce fluorine for various applications

Table 2: Abbreviations, compositions and suppliers for commercial fluorinated polymers*

Abbreviation	Generic polymer name	Manufacturer	Product Tradename
PFA	Perfluoroalkoxy copolymer	DuPont Hoechst Daikin	Teflon®PFA Hostaflon PFA Neoflon
PVDF	Poly(vinylidene fluoride)	Atochem USA Atochem Solvay Daikin Industries Kureha Chemical	Kynar Foraflon Solef Neoflon KF
PCTFE	Poly(chlorotrifluoroethylene)	3M Allied-Signal Atochem Daikin Industries	Kel-F Aclon Voltalef Daiflon
FEP	Fluorinated ethylene-propylene	DuPont Hoechst Montefluos Daikin Industries Ltd	Teflon® FEP Hostaflon Algoflon Neoflon
PTFE	Poly(tetrafluoroethylene)	DuPont ICI Hoechst Montefluos Asahi-ICI Fluoropolymers Daikin Industries Ltd	Teflon® Fluon Hostaflon TFE Algoflon Fluon Polyflon

*does not include Chinese or Russian commercial efforts/products

Costs of fluorinated bulk components, their valuable interfacial properties and their relatively poor bulk performance properties have thus prompted vigorous development of fluorinated coatings and film alternatives as logical niches for these materials. Additionally, processing problems intrinsic to fluorinated polymers, including high crystallinity and poor solubility have limited applications.

Market Issues

World–wide consumption of fluorinated polymers in 1994 was about 70 million kg, with an annual growth rate of 3.2%.(*46*) Approximately 5 million kg of PTFE were consumed as aqueous dispersions to coat cookware, to impregnate fabrics such as fiber glass, and to produce self-lubricating mechanical parts.

The leading market for FEP is the insulation of plenum wire and cable. About 5%, or 500 thousand kg, is used in film applications, such as release films for vacuum bagging during composite molding.

PVDF is a medium priced melt processable fluoropolymer with excellent resistance to weathering, radiation and most chemicals. Approximately one half of the 5 million kg of PVDF sold each year is used as architectural coatings. Dispersion-type resins are applied to metals, such as aluminum and galvanized steel, using coil coating and spray finishing techniques, for use on exterior surfaces.

Poly(vinyl fluoride), PVF, is not melt processable. It is sold almost exclusively in the form of film, fabricated using a gel extrusion process. The excellent toughness, flexibility, and weather resistance of PVF film has led to a variety of uses in the construction and transportation industries, where it is used as a laminate. World-wide sales are about 1.5 million kg per year.

PFA is a high priced melt processable copolymer of TFE and perfluoropropyl vinyl ether. The major use of PFA is in wafer carriers and in pipes and fittings used in the manufacture of semiconductors. PFA is also used in combination with PTFE in formulations for coating cookware.

Poly(chlorotrifluoro-ethylene), PCTFE, is a high priced, high performance, melt-processable thermoplastic used in severe environments. PCTFE exhibits superior creep resistance and resists embrittlement in contact with liquid oxygen, liquid nitrogen and liquified natural gas. It is not affected by ultraviolet (UV) or gamma radiation. World-wide market is about 25 thousand kg per year.

Amorphous Thermoplastic Fluoropolymers. A new class of amorphous thermoplastic fluoropolymer was introduced in 1989. DuPont introduced Teflon® AF, a copolymer of TFE with the cyclic comonomer, perfluoro-(2,2-dimethyl-1,3-dioxole)

(PDD). Asahi Glass introduced Cytop®, a similar copolymer of TFE with a cyclic perfluorinated co-monomer. The current price of these specialty polymers is about ten dollars per gram. Since they are amorphous, they are optically clear and exhibit excellent light transmittance from UV to near infra-red, low refractive index and low dielectric constant. These materials are finding limited applications as optical fiber core and cladding, anti-reflective coatings and interlayer dielectrics.(47-50)

Surface enrichment as a coating strategy

Perfluorinated liquids have long been known to exhibit non-ideal mixing behavior with their aliphatic analogs.(51) This property, combined with the significantly lower solid state surface tension ($\gamma_{s/v}$ or γ_c) of perfluorinated chemistry over hydrocarbon chemistry at an air interface,(22) comprise a substantial thermodynamic driving force for structuring films and surfaces. Exploiting these mixture thermodynamics and their kinetic limits from intrinsic solubilities, mass transfer and partitioning provides some useful advantages for forming perfluorinated films.(52,53) A mixture of hydrocarbons and perfluorocarbons cast together on a surface in air will spontaneously tend to phase separate, 'blooming' the lower interfacial energy components to the air interface and minimizing mixing energies between the two bulk component phases.(52-55) The result then is a stratified film where the perfluorinated species have formed a layer facing the air phase and hydrocarbon mixing is minimized. This surface enrichment is useful to design stratified films of particular surface chemistry.(32-35) Additionally, it holds significant advantages in minimizing the amount of expensive perfluorinated material necessary to create a fluorinated surface: only an overlayer need be created to impart the desired surface properties. This then permits fabrication of high value added surfaces containing fluorine over less expensive commodity material matrices.

Small molecule perfluorinated component films as an alternative to fluorinated polymer coatings.
Fluorinated coating components must not be necessarily restricted to fluorinated polymers, although traditionally, the technology has focused on polymers. Significant commercial interest and markets have been generated from the use of small molecule perfluorinated components as coating materials. The fabric coating product, Scotchgard™, from 3M is one example. Dupont's newly developed coating represents another.(56) In these cases, the coating matrix is an inexpensive base coating material/resin and is doped with small amounts of the more expensive perfluorinated film former. Because of surface migration/enrichment effects described above for low surface-energy perfluorinated components, exposure of this material mixture to air (for example, by spraying or coating from solvent) initiates the formation of the film's stratified matrix. A high value-added surface layer of perfluorinated chemistry is created with time if migration of components toward equilibrium distributions is allowed (e.g., solvent does not 'flash' off or viscosity is not too high). This migration time is less than the cure time of the coating, permitting full maturation of perfluorinated surface properties prior to film cure. Advantages of such a strategy

include use of less-expensive small molecule perfluorinated components having improved solubilities in environmentally friendly propellants (for aerosols) or dispersants (for coatings/paints). Additionally, intrinsically higher small molecule diffusion coefficients permit more rapid surface 'blooming' of perfluorinated components to create an enriched surface overlayer, reducing necessary time to cure.

Small perfluorinated molecule film forming components include perfluoroalkanoic acids, perfluorinated amines, perfluoroethers, perfluoroalkylolefins and perfluoroalkylsulfonamides. Often, these components have low surface tensions in addition to carrying functional groups to surfaces.(8) Additionally, several cases exist where perfluorinated monomers (e.g., acrylates, epoxides) mixed with aliphatic resins readily form enriched stratified overlayers that can then be polymerized post-bloom.(54,55)

Fluorinated plasma deposition

In addition to more conventional solution-based methods to produce fluorinated coatings, gas-phase deposition of fluorinated precursors can proceed to confluent, robust films via plasma polymerization processes.(57-59) Volatile small molecule feed gases exposed to continuous wave or pulsed microwave energy produce complex gas phase radical and ion chemistry resulting in surface-induced solid phase film quenching and film deposition. This rather well-developed method of surface modification can be applied where solution coating is not technically feasible. Advantages to this technique include production of conformal and usually defect-free films in a one-step process. Line-of-sight coating is not reliable. Disadvantages of the method surround the requirement for volatile gas-phase perfluorocomponents, resulting crosslinked film product, lack of control of aspects of film chemistry, resultant film surface roughness in some cases and low deposition rates and throughput. Fluorinated plasmas have received considerable scientific and technical attention and continue to be developed for industrial applications, particularly for microelectonics applications as low-dielectric materials.

Future needs

Despite the relatively flat current market in fluorochemicals, applications and new markets for fluorinated coatings are substantial. To realize this potential, various dimensions of fluorinated films and coatings technology require further development. Attention might well be paid to fabricating fluorinated materials for coatings and films with:

- Increased radiation resilience.
  Relatively poor radiation stability of fluorinated materials compromises their use in aerospace and many outdoor technical and recreation markets. Loss of water repellency for fabric coatings is one classic example. Improving these materials'

capabilities to withstand even ambient UV radiation or protecting this intrinsic weakness with formulations/additives is one possibility.

- Improved aftermarket coating applications.
  Many fluorinated coatings must be applied to laboratory-grade clean or virgin surfaces. Further aftermarket applications of protective fluorinated coatings to surfaces after significant environmental exposure or soiling has not produced favorable results, despite routine field (not laboratory) surface cleaning procedures. Problems surround the cost of effective field application methods using non-benign solvents, plus the lack of durability of films coated onto contaminated surfaces.

- Reduced bulk materials cost.
  Perfluorinated precursors have always been an expensive specialty chemical. This is likely not going to change unless new methods to make these materials are developed. *In situ* fluorination and other selective, direct hydrocarbon transformations to perfluorinated analogs represent attempts to accomplish this. Fluorinated materials will always represent a 'gold standard' unless breakthroughs are made to reduce their cost. Traditional ostracism of fluorine chemistry away from the organic chemist's domain is unnecessary and counterproductive.

- Improved perfluorinated materials processing.
  Post-synthetic processing methods that side-step currently costly procedures necessary for PTFE and similar polymers suffering poor processability are another logical target. Typically, because of lack of solvents accorded to its high crystallinity, PTFE must be sputtered by ablation from solid targets or sintered from emulsions to form films and coatings. These films then are often rough and porous. PTFE solid blocks are often made by sintering PTFE bead suspensions at high temperature and pressure. Mechanical and machining properties of such materials are sub-standard. Most effective approaches to improvement appear to be focused on synthesis of new perfluorinated materials that, by their nature, avoid costly processing pathways (e.g., lower crystallinity, improved solubility). Another strategy should focus on optimizing blooming effects for minority perfluorinated components to form overlayers on less expensive hydrocarbon resins. While blooming is frequently an industrial trade secret, few systematic studies have been published to fully explore the process and possibilities. One might ask "at what bulk composition is a perfluorinated surface no longer a perfluorinated surface?" The answer ultimately lies in proven performance in a given application. However, scientifically, perfluorinated chemistry and complete overlayer formation, yielding a fluorinated film by definition, can be detected using modern surface analytical methods (e.g., XPS or ToF-SIMS) for bulk perfluorochemistry contents below 1wt% in hydrocarbon resins. This means that sufficiently little perfluorinated component might be required for a desired surface performance to make the entire

material process economically attractive if blooming and surface enrichment effects could be optimized.

- Possible alternatives to perfluorinated surface chemistry that produce the same effects.

Chemical and physical principles behind the superior properties of perfluorocarbon species as films are generally accepted anecdotally or empirically; few of them are actually explained from first principles. Much hand-waving exists in attempting to directly correlate the structural and physical chemical aspects of the C-F bond to perfluoroalkyl structure and further higher order collective properties observed for these materials. Direct evidence as to why a C-F bond (with associated higher dipole moment than a C-H bond) is in fact not attracted to polar liquids, including water, remains enigmatic. Historical consensus dating to Zisman points to a need to maximize the amount of surface-exposed -CF$_3$ groups compared to other chemistries, including -CF$_2$- groups, in order to optimize perfluorinated surface properties. Maximizing surface fluorine content has been the engineering result of this analysis. Perfluorinated component surface blooming has been one approach to achieve this experimentally. Orientation of surface perfluoroalkyl chains to expose -CF$_3$ chain ends and bury other chemistry has been a traditional focus. Helical perfluoroalkyl chains occupy one and one-half times the volume of an all-trans hydrocarbon chain (28Å^2/chain vs. 18Å^2/chain, respectively (*25,60*)). Additionally, a typical R$_f$ chain available (e.g., 10 perfluorinated carbon units) comprises only one full helical turn (if assuming a PTFE-type helical structure (*21*)). Such a structure standing on end might be considered as a stiff, short stump. Spontaneous surface orientation of these R$_f$ stumps for structural self-avoidance and surface packing purposes is a logical consequence. Hence, many studies examining natural orientation and surface structural anisotropy in perfluorinated chain films have found it. Locking such orientations into place using polymerization has followed. Many commercial perfluoroalkylated monomers offer this possibility. Blooming these monomers to surfaces of aliphatic resins with subsequent polymerization is a promising strategy.

Another possibility remaining relatively unexplored involves surface topological and lateral distribution of perfluorinated chemistry to affect performance. Films to date usually assume "more is better" when it comes to enriching perfluorinated surfaces. However, recent work has shown that super non-wetting properties are achieved for non-fluorinated surfaces of fractal surface topology (*61*). Perfect presentation of perfluorinated chemistry may therefore not be such an issue *per se* if topology of perfluorinated films is modulated in a predictable and controlled fashion. Additionally, lateral surface distributions of perfluorinated chemical domains on the size order required to modulate properties such as tribology, adhesion and wetting have not been explored. It is conceivable that total surface coverage with perfluorinated chemistry is not in fact either essential or optimal when lateral control of perfluorinated species density, orientation, topology and distribution are used in tandem in the film fabrication process.

Acknowledgements. DWG wishes to acknowledge support from NSF grant DMR-9596023, a 3M Faculty Fellowship and a Dupont Research Award that have allowed research in various physical and chemical aspects of fluorinated films and coatings at Colorado State University. Additionally, he recognizes support of the Japanese Society for Promotion of Science (JSPS) and MITI agencies through an invited joint NSF-JSPS professorship in Prof. Teruo Okano's laboratory (Tokyo Women's Medical University, Japan) that permitted the time necessary to write this chapter and edit this book.

Literature cited.

1. Plunkett, R.J., US Pat. 2,230,654 **1941.**
2. Feiring, A., in *Organofluorine Chemistry: Principles and Commercial Applications*, Banks, R.E., Tatlow, J.C., Smart, B.E., eds., Plenum Press, New York, NY, 1994, Chapter 15.
3. Banks, R.E., *Fluorocarbons and Their Derivatives*, 2nd ed., Macdonald, London, 1970.
4. Banks, R.E., J.C. Tatlow, J.C., *J.Fluorine Chem.*, **1986** *33* 71.
5. Banks, R.E., Tatlow, J.C., in *Organofluorine Chemistry: Principles and Commercial Applications*, Banks, R.E., Tatlow, J.C., Smart, B.E., eds., Plenum Press, New York, NY, 1994, Chapter 1.
6. Powell, R.L., Steven, J.H., in *Organofluorine Chemistry: Principles and Commercial Applications*, Banks, R.E., Tatlow, J.C., Smart, B.E., eds., Plenum Press, New York, NY, 1994, Chapter 28.
7. Simons, J.H., Block, L.P., *J. Am Chem. Soc.*, **1939** *61* 2962.
8. Smart, B.E., in *Organofluorine Chemistry: Principles and Commercial Applications*, Banks, R.E., Tatlow, J.C., Smart, B.E., eds., Plenum Press, New York, NY, 1994, Chapter 3.
9. Williams, D.E., Houpt, D.J., *Acta Cryst.*, **1986** *B42* 286.
10. Hansch, C., Leo, A., *Substituent Constants for Correlation Analysis in Chemistry and Biology*, Wiley, New York, NY, 1979.
11. Thornber, C.W., *Chem. Soc. Rev.,* **1979** *8* 563.
12. Bott, G., Field, L.D., Sternhill, S., *Austral. J. Chem.*, **1987** *40* 35.
13. Russell, T.P., Rabolt, J.F., Twieg, R.J., Siemens, R.L., Farmer, B.L., *Macromolecules* **1986** *19* 1135.
14. Twieg, R.J., Russell, T.P., Siemens, R., Rabolt, J.F., *Macromolecules* **1985** *18* 1361.
15. Zhang, W.P., Dorset, D.L., *Macromolecules* **1990** *23* 4322.
16. Bunn, C.W., Howells, E.R., *Nature* **1954** *174* 549.
17. Clark, E.S., Muus, L.T., *Z. Kristallogr.* **1962** *117* 119.
18. Naselli, C., Swalen, J.D., Rabolt,J.F., *J. Chem. Phys.*, **1989** *90* 3855.

19. Schneider, J., Erdelen, C., Ringsdorf, H., Rabolt, J.F., *Macromolecules* **1989** *22* 3475.
20. Tsao, M.-W., Hoffman, C.L., Rabolt, J.F., Johnson, H.E., Castner, D.G., Erdelen, C., Ringsdorf, H., *Langmuir* **1997** *13* 4317.
21. Clark, E.S., *Polymer* **1999** *40* 4659.
22. Kissa, E., *Fluorinated Surfactants: Synthesis-Properties-Applications*, Marcel Dekker, New York, 1994.
23. Pittman, A.G., in *Fluoropolymers*, LA Wall, ed., Wiley, New York, NY, 1972, Chapter 13.
24. Hare, E.F., Shafrin, E.G., Zisman, W.A., *J. Colloid Sci.*, **1954** *58* 236.
25. WA Zisman, in *Contact Angle Wettability, and Adhesion*, Gould, RF, ed., ACS Adv. Chem. Ser. P43, Washington, DC 1964.
26. Johnson, R.E., Jr., Dettre, R.H., in *Wettability*, Berg, J.C. ed, Marcel Dekker, New York, NY, 1993.
27. Stone, M., Nevell, T.G., Tsibouklis, J., *Mater. Lett.*, **1998** *37* 102; Hopken, J. Sheiko, S., Czech, J., Moller, M., *Am. Chem. Soc. Div. Polym. Chem. Prepr.* **1992** *33* 937.
28. Zhang, Y.X., Da, A.-H., Hogen-Esch, T.E., *J. Polym. Sci., Polym. Lett.*, **1990** *28* 213.
29. Bar, G., Thomann, Y., Brandsch, R., Cantow, H.-J., Whangbo, M.-H., *Langmuir*, **1997** *13* 3807.
30. Kobayashi, H., Owen, M.J., *Trends Polym. Sci.*, **1995** *3* 330.
31. Doeff, M.M., Lindler, E., *Macromolecules*, **1989** *22* 2951.
32. Sun, F., Mao, G., Grainger, D.W., Castner, D.G., *Thin Solid Films*, **1994** *242* 106.
33. Sun, F., Grainger, D.W., Castner, D.G., Leach-Scampavia, D.K., *Macromolecules*, **1994** *27* 3053.
34. Sun, F., Castner, D.G., Mao, G., Wang, W., McKeown, P., Grainger, D.W., *J. Am. Chem. Soc.*, **1996** *118* 1856.
35. Wang, W., Castner, D.G., Grainger, D.W., *Supramolec. Sci.*, **1997** *4* 83.
36. Kim, D.K., Lee, S.-B., Doh, K.-S., *J. Colloid Interface Sci.*, **1998** *205* 417.
37. Anton, R.D., Darmon, M.J., Graham, W.F., Thomas, R.R., US patent 5605956 (1997).
38. Winnik, M.A., *J. Coatings Technol.*, **1996** *68* 39.
39. Moore, G., Zhu, D.W., Clark, G., Pellerite, M., Burton, C., Schmidt, D., Coburn, C., *Surface Coatings Int'l* **1995** 377.
40. *Modern Fluoropolymers*, J. Schiers, Wiley, Chichester, 1977.
41. Hale, A., *Am. Chem. Soc. Div. Polym. Chem. Prepr.* **1998** *39* 977.
42. Inukai, H., Yasuhara, T., Kitahara, T., PCT Intl. Appl. WO 89 05,287 (1989) [CA 111, 195615 (1989)].
43. Bliefert, C., Boldhaus, H.-M., Erdt, F., Hoffmann, M., *Kunstoffe*, **1986** *76* 235.
44. Chiao, C.C., U.S. Pat. 4,828,585, **1989** to Dow [*CA* 111, 176891 (1989)].
45. Belaish, I., Davidov, D., Selig, H., McLean, M.R., Dalton, L., *Angew. Chem., Int. Ed. Engl.*, **1989** *28* 1569.

46. Haley, M.J., Leder, A., Sakuma, S., *CEH Marketing Research Report: Fluoropolymers*, Chemical Economics Handbook, SRI International, Palo Alto, CA 1995.

47. Murarka, S.P., *Solid State Technol.*, **1995**(3) 83.

48. Ting, C.H., Seidel, T.E., *Mat. Res. Soc. Symp.* **1995** *381* 3.

49. Feiring, A.E., Auman, B.C., Wonchoba, E.R., *Macromolecules* **1993** *26* 2779.

50. Singer, P., *Semiconductor Int'l* **1996** *19* 88.

51. Scott, R.L., *J. Phys. Chem.* **1958** *62* 136.

52. Thomas, R.R., Anton, D.R., Graham, W.F., Darmon, M.J., Sauer, B.B., Stika, K.M., Swartzfager, D.G., *Macromolecules*, **1997** *30* 2883.

53. Thomas, R.R., Anton, D.R., in *Fluorinated Surfaces Coatings and Films*, ACS Symp. Ser., ed., Castner, DG, Grainger, DW, ACS Press, Washington, D.C., 1999.

54. Schnurer, A.U., Holcomb, N.R., Gard, G.L., Castner, D.G., Grainger, D.W., *Chem. Mater.* **1996** *8* 1475.

55. Winter, R., Dixon, P., Gard, G.L., Castner D.G., Holcomb, N.R., Hu Y.-H., Grainger, D.W., *Chem. Mater.* **1999,** *11* 3044.

56. Sauer, B.B., McLean, R.S., Thomas, R.R., Langmuir **1998** *14* 3045.

57. D'Agostino, R., ed., *Plasma Deposition, Treatment and Etching of Polymers*, Academic Press, San Diego, CA, 1990.

58. Yasuda, H., *Plasma Polymerization*, Academic Press, New York, NY, 1985.

59. Butoi, C.I., Mackie, N.M., Barnd, J.L., Fisher, E.R., Gamble, L.J., Castner, D.G., *Chem. Mater.* **1999** *11* 862.

60. Bernett, M.K., Zisman, W.A., *J. Phys. Chem.*, **1963** *67* 1534.

61. Onda, T., Shibuichi, S., Satoh, N., Tsujii, K, *Langmuir* **1996** *12* 2125.

Chapter 2

AFM Study on Lattice Orientation and Tribology of SAMS of Fluorinated Thiols and Disulfides on Au(111): The Influence of the Molecular Structure

Holger Schönherr and G. Julius Vancso

Faculty of Chemical Technology, Polymer Materials Science and Technology, University of Twente, P.O. Box 217, 7500 AE Enschede, The Netherlands

The molecular packing and the tribological properties of fluorinated thiols and disulfides in self-assembled monolayers (SAMs) on Au(111) were studied. Atomic force microscopy (AFM) images with molecular resolution of the tail groups unveiled the lattice structure. The relative orientation of the tail group lattices with respect to the underlying Au(111) could be determined by the comparison of the lattice orientation with the orientation of the triangular Au(111) terraces. Depending on the molecular structure, a transition from a p (2×2) structure for short chain molecules to a c (7×7) structure for long chain molecules was observed for three homologous series of perfluoro organosulfur compounds. The friction force measured with silicon nitride tips was found to depend on the chain length of the molecules. Pull-off forces and adhesion hysteresis in AFM experiments did not correlate with the magnitude of friction. The observed qualitative correlation of disorder and increase in friction force supports the friction mechanism of energy dissipation in SAMs.

Introduction

Self-assembled monolayers (SAMs) of organic molecules on solid supports are considered as valuable model systems for interfacial studies on e.g. wettability, adhesion, lubrication, tribology etc. (1, 2). Especially SAMs of organosulfur-based compounds (thiols, disulfides and sulfides) on Au are widely used due (a) to the large number of different functional groups that can be introduced in the ω-position of the alkane chain and (b) the high degree of organization in the SAMs.

Scanning probe microscopy's such as scanning tunneling microscopy (STM) (3) and atomic force microscopy (AFM) (4) have proved to be suitable methods to investigate the lattice structure of SAMs of e.g. alkane thiols. The orientation of the tail group lattice of the adsorbed molecules was shown by various methods to be often correlated with the lattice orientation of the underlying Au(111) substrate. Examples include the commensurate ($\sqrt{3} \times \sqrt{3}$) R 30° lattice of ordinary n-alkane thiols (a) (5, 6), disulfides (b) (7) and thioethers (8), as well as the lattices formed by fluorinated thiols (c) (9) or fluorinated disulfides (d) (10) on Au(111). For all these compounds the nearest neighbor direction in the lattice is rotated by 30° with respect to the underlying Au(111).

The orientation of the tail group lattice is related to the interaction of the sulfur atoms with the Au(111) surface. The sulfur atoms are believed to bind to the three-fold hollow sites of the Au(111) and form a ($\sqrt{3} \times \sqrt{3}$) R 30° lattice (11, 12). The alkane chains form a similar commensurate tail group lattice (13). However, there are other factors that contribute to the lattice structure such as chain-chain interaction, physical size of terminal substituents or the shape of the cross section of the molecule (1). Thus, it is not surprising to find a number of reports on high order commensurate or even incommensurate lattices formed by various classes of thiols and disulfides, e.g. fluorocarbon thiols (9) and disulfides (10), azobenzene terminated thiols and disulfides (14), as well as triphenylene based thiols (15). The balance of forces seems to be in favour of a tail group dominated structure in these cases.

Surprisingly, fluorinated thiols and disulfides do not form a p (2 × 2) structure (16) but a high order commensurate c (7 × 7) structure (Figure 1). In general, fluorocarbon chains adopt a helical conformation (17). Compared to an n-alkane chain a fluorocarbon chain possesses a higher torsion energy, thus it can be regarded as a rigid rod (18). Recently, the structure of the molecules was successfully correlated with the symmetry of the tail group lattices. The transition from oblique to hexagonal symmetry of the lattices for various thiols and disulfides was reported by Nelles et al. (19). Furthermore, a simple model that allows one to correlate molecular structure, especially chain length, and lattice symmetry was proposed (19). In an earlier AFM investigation of the lattice structure of fluorinated thiols and disulfides on sputtered gold we showed that the fluorocarbon segment is responsible for the observed lattice spacing of 5.8 ± 0.2 Å (20). Unlike for similar hydrocarbon disulfides and thiols (10, 21), the fluorocarbon molecules always form a hexagonal lattice, despite significant structural differences such as e.g. the presence of ester bonds within the chain.

In the present paper we discuss AFM results on SAMs of fluorinated thiols and disulfides on Au(111). In particular, the relation of molecular structure of the adsorbate and the relative orientation of the tail group lattice with respect to the orientation of the underlying Au(111) lattice and the tribological properties of the SAMs are addressed. Both the orientation of the lattices and the tribological properties were found to depend on the structure of the molecules, especially on the chain length and on intermolecular interactions. The results of the friction

17

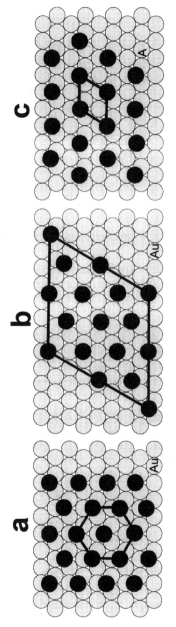

Figure 1. Tail group lattice structures of SAMs on Au(111): (a) ($\sqrt{3} \times \sqrt{3}$) R 30°; (b) c (7 × 7); (c) p (2 × 2) (The diameter of the tail groups is not to scale.)

measurements are in full agreement with the concept of energy dissipation as dominant friction mechanism in these films (*22, 23*).

Experimental

Substrate and SAM Preparation. Evaporated gold substrates (borosilicate glass, 2 nm Cr, 250 nm Au) were purchased from Metallhandel Schröer, Lienen (Germany). Au(111) samples were prepared by flame annealing in a high purity hydrogen flame for ca. 10 minutes. The substrates showed numerous Au(111) terraces which were usually several μm large. On the annealed substrates frequently equilateral triangular terraces were found (Figure 2). Monolayers were prepared by self-assembly from 0.5 to 1.0 mM solutions of the corresponding thiol or disulfide in dichloromethane or ethanol (p.a., Merck) as described previously (*21*). The compounds were synthesized as described in references 20 and 21.

Atomic Force Microscopy (AFM). We used commercial NanoScope II and NanoScope III Multimode force microscopes (Digital Instruments (DI), Santa Barbara, Cal., USA) equipped with 1 and 10 μm scanners. For lattice resolution imaging, cantilevers (DI) with nominal spring constants of 0.06, 0.12 and 0.38 N/m were used. For imaging with minimized normal forces ethanol is usually used as a medium (using a liquid cell) (*24*). However, in order to exclude possible albeit unlikely rearrangements of the lattice structure due to interactions between SAMs and solvent, measurements reported in this work were performed exclusively in air. Friction measurements were carried out in a home-built glass chamber which was purged with dry nitrogen. Triangular silicon nitride cantilevers with integrated tips (DI) (nominal spring constant 0.38 N/m) were used. Before the experiments the AFM set-up was equilibrated for at least 5 hours in order to minimize the thermal drift. Before and after each measurement freshly cleaved mica was scanned in order to record a reference friction curve. The friction data was obtained by scanning the sample in 90° to the long axis of the cantilever. 200 nm line scans were performed on SAMs on flat triangular Au(111) terraces at a scan rate of 5 Hz (scan velocity ≈ 1.0 μm/s) for variable external load. The friction force was calculated from the so-called "friction loops" (*25*). The data was averaged over at least three sets of experiments performed with the same tip / cantilever on different terraces. The lateral force constant was estimated to be 87 N/m using the double beam approach, based on scanning electron microscopy (SEM) images of the cantilever dimensions and the known elastic constant of the material, as described in reference 26. A friction signal of 0.01 V corresponds to a friction force of 2.7 nN. As the calibration procedure is not very accurate, the results were stated as friction signal (photodiode output). It is accurate to compare results obtained on different SAMs with the same set-up as all essential parameters (cantilever, laser adjustment etc.) (*27*) were kept constant. The *radii* of the tips used in the friction measurements were measured with a scanning electron microscope (Jeol JSM-T 220A) at a voltage of 20 kV. All the tip *radii* were found to be in the range of 75 ± 25 nm. However, in order to avoid artefacts due to

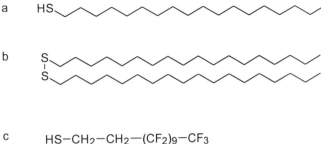

a   HS∼∼∼∼∼∼∼∼∼∼∼∼

b   S∼∼∼∼∼∼∼∼∼∼∼∼
     |
     S∼∼∼∼∼∼∼∼∼∼∼∼

c   HS–CH₂–CH₂—(CF₂)₉–CF₃

d   S—(CH₂)₂–O–CO—(CH₂)₂—(CF₂)₇–CF₃
     |
     S—(CH₂)₂–O–CO—(CH₂)₂—(CF₂)₇–CF₃

Chart 1. *The lattice structure of the thiols and disulfides shown were successfully determined using AFM. n-alkane thiols and disulfides, such as **a** and **b**, form commensurate ($\sqrt{3} \times \sqrt{3}$) R 30° structures (Figure 1a) (6), while the fluorinated compounds **c** and **d** form a high order commensurate c (7 × 7) structure (Figure 1b) (9, 10). For all structures the nearest neighbor direction in the tail group lattice is rotated by 30° with respect to Au(111).*

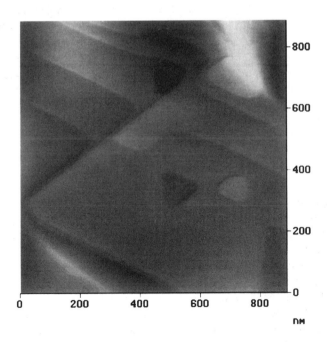

Figure 2. *AFM image of Au(111) terraces (z-scale 3nm).*

different tip *radii* and spring constants, the results obtained with a single cantilever / tip assembly are presented for each series of SAMs.

Results and Discussion

The AFM studies were carried out on triangular Au(111) terraces as shown in Figure 2. These substrates provide sufficiently well ordered and well characterized Au(111) (*28*), and they are atomically smooth except for monoatomic steps and holes (etch pits) (*29*). Therefore high resolution lattice imaging as well as quantitative friction measurements are not obscured by any influence of the substrate morphology.

Tail Group Lattice Orientation. The tail group lattices of the SAMs of the compounds listed in Chart 2 were imaged with molecular (lattice) resolution on triangular Au(111) terraces (Figure 2). The terrace edges correspond to nearest neighbor directions of the Au(111) lattice. As the terrace edges can be easily imaged at low magnification, the orientation of the tail group lattices (Figure 1) could be determined accurately with respect to the gold lattice in a non-destructive manner. As reference, we imaged the well-known and -characterized SAMs of n-alkane thiols with chain lengths between 10 and 18, as well as one fluorinated thiol (compound 1).

All fluorinated compounds investigated in this study formed a hexagonal lattice with a lattice constant of 5.7 - 5.8 Å. Surprisingly, the orientation of the lattice with respect to Au(111) was found to vary with the molecular structure of the adsorbate molecules, and in several cases the orientation was found to depend even on the chain length.

An unprocessed AFM image obtained on a SAM of fluorinated thiol 1 is shown in Figure 3. The hexagonal lattice is well resolved. In addition, the thermal drift of the AFM scanner is minimized as one can conclude from the negligibly small deviation of the position of corresponding tail groups from the trajectories of the lattice. The lattice constant, as well as the lattice orientation with respect to Au(111) agree with the results published in the literature (*9*).

The relative orientation of the tail group lattice of the fluorinated aromatic thiol 11 was found to be parallel to the terrace edges of the Au(111) and hence with respect to the Au(111) lattice. Thus, a p (2 × 2) structure is formed. To our knowledge this is the first observation of this tail group lattice structure (Figure 4).

For the amide thiols **6-8**, a change in the length of the $(CF_2)_n$ segment caused the nearest neighbor direction to rotate with respect to Au(111), i.e. the lattice changed from a p (2 × 2) to a c (7 × 7) structure. An example of the c (7 × 7) structure formed by the tail groups of thiol 7 on Au(111) is represented in Figure 5. The image was obtained with minimized forces (*20*).

For the disulfides **3-5** the change in $(CH_2)_n$ segment length caused a similar rotation. When adding a single methylene group to the hydrocarbon spacer in compound **10**, a similar relative rotation of the lattice is observed (compound **9**). The results are summarized in Table 1.

1 $HS-CH_2-CH_2-(CF_2)_9-CF_3$

2 $S-(CH_2)_2-O-CO-(CH_2)_2-(CF_2)_7-CF_3$
   $|$
   $S-(CH_2)_2-O-CO-(CH_2)_2-(CF_2)_7-CF_3$

3 -5 $S-\ldots\,_2)_n-O-CO-(CF_2)_8-CF_3$
   $|$
   $S-(CH_2)_n-O-CO-(CF_2)_8-CF_3$ $n = 2, 6, 11$

6 -8 $HS-CH_2-CH_2-NH-CO-(CF_2)_n-CF_3$ $n = 6, 7, 8$

9 $S-(CH_2)_2-CO-O-(CH_2)_2-(CF_2)_7-CF_3$
   $|$
   $S-(CH_2)_2-CO-O-(CH_2)_2-(CF_2)_7-CF_3$

10 $S-(CH_2)_2-CO-O-CH_2-(CF_2)_5-CF_3$
   $|$
   $S-(CH_2)_2-CO-O-CH_2-(CF_2)_5-CF_3$

11 $HS-(CH_2)_4-O-\langle\!\!\!\bigcirc\!\!\!\rangle-S-CH_2-(CF_2)_9-CF_3$

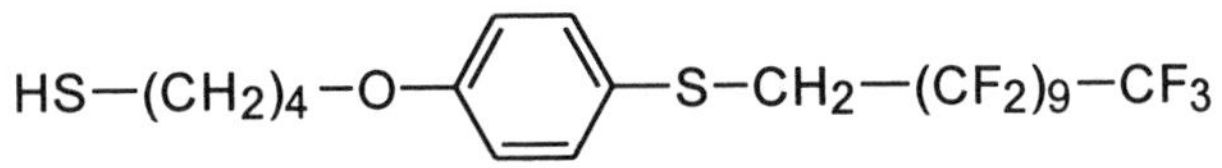

Chart 2. *Compounds investigated in this study.*

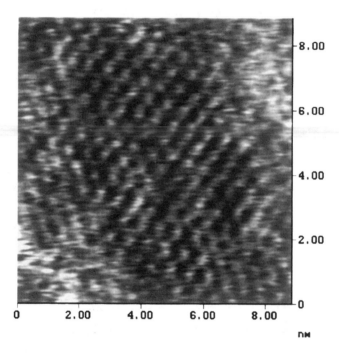

Figure 3. Unprocessed AFM image of fluorinated thiol *1* (reference) on Au(111).

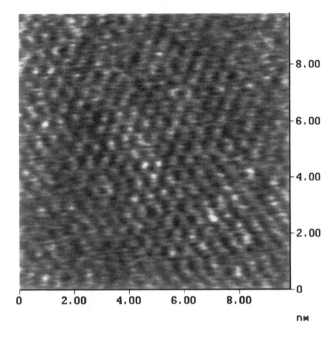

Figure 4. Unprocessed AFM image of fluorinated thiol 11 on Au(111).

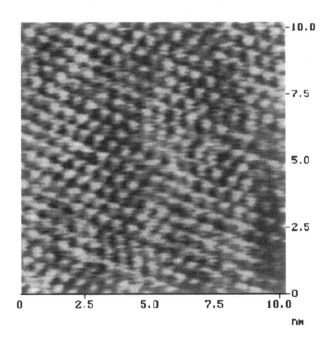

Figure 5. Unprocessed AFM image of fluorinated thiol 7 on Au(111).

Table 1. Lattice constants, lattice orientation of tail group lattice, and closest commensurate tail group lattice.

compound	lattice constant Å	rotation of NN direction with respect to Au(111)	closest commensurate tail group lattice
1	5.8 ± 0.2	30°	c (7 × 7)
2	5.8 ± 0.2	30°	c (7 × 7)
3	5.7 ± 0.2	0°	p (2 × 2)
4	5.7 ± 0.2	0°	p (2 × 2)
5	5.7 ± 0.2	30°	c (7 × 7)
6	5.8 ± 0.3	0°	p (2 × 2)
7	5.7 ± 0.2	30°	c (7 × 7)
8	5.7 ± 0.2	30°	c (7 × 7)
9	5.8 ± 0.2	30°	c (7 × 7)
10	5.8 ± 0.2	0°	p (2 × 2)
11	5.7 ± 0.1	0°	p (2 × 2)
n-alkane thiols (n = 10, 12, 16, 18)	5.0 ± 0.2	30°	($\sqrt{3} \times \sqrt{3}$) R 30°

The factors that govern the relative orientation of the tail group lattice with respect to the underlying Au(111) substrate are difficult to quantify. As stated previously, there are several contributions to the stabilization in SAMs: (1) binding of the sulfur to the gold, (2) the tail-tail interaction, and (3) the interactions of the terminal groups (*1*). In addition, there is evidence for "through space" interaction e.g. between metal substrate and liquids used in wetting experiments (*30*). Furthermore, the anchoring properties of liquid crystals on SAMs show a delicate balance of the different forces. Specifically, the wetting behavior of SAMs of disulfides 3 and 5 was found to be identical with water, while the anchoring of nematic liquid crystals of 5-cyanobiphenyl is different. On SAMs of 3 planar anchoring is found and on SAMs of 5 homeotropic anchoring is present, respectively (*31*). Therefore a preference for a certain tail group lattice orientation with respect to Au(111) does not seem unlikely. Calculations of lattice energies and molecular modelling are necessary to quantify the relative contributions of the different interactions.

The observed chain length-dependent transition from a p (2 × 2) to a c (7 × 7) structure seems to be systematic with increasing length of the molecule. The results for the amide thiols 6 - 8 rule out the possibility of an odd-even effect.

Tribological Properties. The friction of partially fluorinated disulfide SAMs (3 - 5) was found to depend on the length of the $(CH_2)_n$ segment. (Figure 6). The $(CH_2)_n$ segment in the lower part of the SAM must be disordered since the diameter of the helical $(CF_2)_n$ segments (5.8 Å) is much larger than the interchain distance typically found for crystalline $(CH_2)_n$ segments (5.0 Å with 30° tilt angle, 4.2 Å for 0° tilt angle). Additionally, the disulfide ester with 10 methylene units is too short to give a SAM with significant proportion of all-*trans* conformation in the hydrocarbon tails (*21*). Therefore the disordered $(CH_2)_n$ segment is very likely responsible for the observed increase in friction with increasing $(CH_2)_n$ segment length as the number of methylene groups, which can dissipate energy, is increasing. Compound 4 with 6 methylene units shows some scatter in the data points and has been omitted from Figure 6 for clarity. However, the magnitude of friction was found to decrease with decreasing number of methylene units ($(CH_2)_{11} > (CH_2)_6 > (CH_2)_2$).

The magnitude of friction force between silicon nitride and SAMs of amide thiols 6 - 8 was found to depend on the length of the fluorocarbon segment (Figure 7). For n = 6 to n = 8 a decrease in friction for increasing $(CF_2)_n$ chain length was observed. These results indicate that the packing of the $(CF_2)_n$ helical segments becomes more stable, or the segments more rigid, with increasing segment length. Thus, less energy can be dissipated by vibrational or rotational modes (*22, 23*).

Preliminary results which we obtained on fluorinated amide thiols with even longer $(CF_2)_n$ segment (n = 10, 12) indicate that the friction increases as compared to n = 8 which would correspond to a minimum in friction as a function of chain length. These observations are in full agreement with the disorder detected in long fluorinated chains by Rabolt et al. (*32*) and with the interpretation of the data presented in this paper (*vide supra*). These results also support our earlier report on the different tip penetration depths into SAMs of these thiols during force-dependent imaging with AFM (*20*).

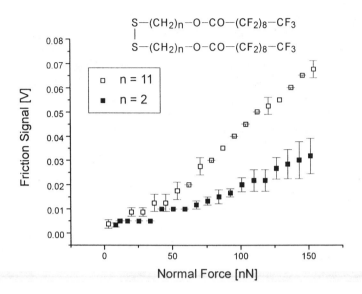

Figure 6. *Friction vs. normal force for different $(CH_2)_n$ segment lengths in disulfides (3, 5).*

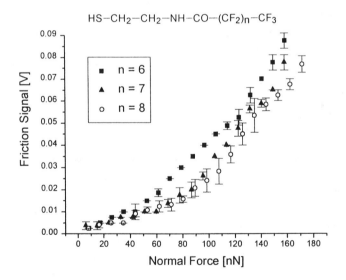

Figure 7. *Friction vs. normal force for different $(CF_2)_n$ segment lengths in amide thiols (6 - 8).*

A third set of friction measurements shows the influence of internal functional groups and the corresponding intermolecular interactions in SAMs of $(CF_2)_n$ thiols with a long terminal perfluorocarbon segment. The wetting properties with water are similar for SAMs of all three thiols (3, 7, 11). In addition to the van der Waals and (unfavourable) dipol-dipol interactions, the molecules in SAMs of thiol 3 (ester), 7 (amide) and 11 (aromatic system) interact with no additional (3), H-bonding (7) (33), or π-π interactions (11). The attractive interactions are expected to increase the rigidity of the SAM as a whole. Thus, the collective friction response is expected to decrease in the order 7 < 11 < 3.

The observed friction data corresponds to this interpretation as the magnitude of friction at the same load increases from the amide via the aromatic system to the ester. Our results clearly indicate that the structural variations of the adsorbate molecules effect the friction behavior as measured by AFM. The terminal functional groups and segments however, are in all cases the same. The observed pull-off forces as a measure for adhesion (friction force is often related to the pull-off forces / adhesion, more specifically the adhesion hysteresis) (34) are virtually the same. Based on these considerations the observed friction behavior can be explained by energy dissipation in the SAM as proposed by Salmeron and co-workers (22, 23).

Conclusions

The molecular structure of the adsorbate molecules in self-assembled monolayers (SAMs) of fluorinated thiols and fluorinated disulfides on Au(111) was shown by atomic force microscopy (AFM) to effect the relative orientation of the tail group lattice with respect to the orientation of the underlying Au(111) lattice. The tribological properties of the SAMs also depended on the structure of the molecules. For a fluorocarbon thiol containing an aromatic group a p (2 × 2) tail group lattice was found whereas for two other classes of compounds (perfluoro amide thiols and bis-perfluoroalkyl ester disulfides) the orientation was found to depend on the chain length of the molecule. The lattices can be described as p (2 × 2) for short chain molecules and c (7 × 7) for the long chain molecules, which is equivalent to a nearest neighbor direction parallel to the Au(111), and at 30° direction respectively. This is to our knowledge the first report of a p (2 × 2) tail group lattice structure in fluorinated SAMs. In a homologous series of bis-perfluoroalkyl ester disulfides the magnitude of the friction force was found to depend systematically on the chain lengths of the molecules. For fluorinated amide thiols a minimum for the friction was observed as a function of perfluorocarbon segment length. In general, increased disorder in the SAMs can be related to increase in friction. No systematic correlation between the magnitude of friction and pull-off forces (or adhesion hysteresis) was observed. The correlation between structural disorder and increase in observed friction force supports the friction mechanism of energy dissipation in SAMs by vibrational and rotational modes proposed by Lio, Charych, and Salmeron (*J. Phys. Chem. B* **1997**, *101*, 380.).

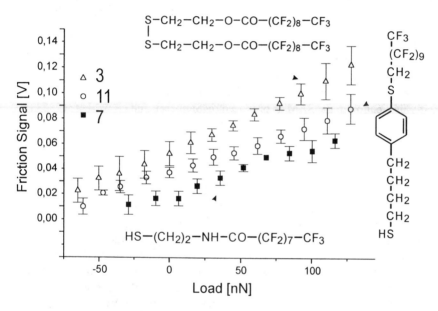

Figure 8. Friction vs. load plot for SAMs of thiols **3, 7, 11**.

Acknowledgement

The authors thank Steven D. Evans and Ban Al-Khairalla for the sample of thiol **11**. This work has been supported by the Council for Chemical Sciences of the Netherlands Organization for Scientific Research (CW-NWO) in the priority program materials (PPM).

References

1. Ulman, A. *Introduction to Ultrathin Films,* Academic Press: Boston, **1991**.
2. Dubois, L. H.; Nuzzo, R. G. *Annu. Rev. Phys. Chem.* **1992**, *43*, 437.
3. Widrig, C. A.; Alves, C. A.; Porter, M. D. *J. Am. Chem. Soc.* **1991**, *113*, 2805. Alves, C. A.; Smith, E. L.; Porter, M. D. *J. Am. Chem. Soc.* **1992**, *114*, 1222.
5. The ($\sqrt{3} \times \sqrt{3}$) R 30° lattice structure was proved by electron diffraction (Chidsey, C. E. D.; Loiacono, D. N. *Langmuir* **1990**, *6*, 682) and He diffraction at low temperatures (Camillone III, N.; Chidsey, C. E. D.; Liu, G.-y.; Putvinski, T. M.; Scoles, G. *J. Chem. Phys.* **1991**, *94*, 8493).
6. Liu, G.-Y.; Salmeron, M. *Langmuir* **1994**, *10*, 367.
7. Strong, L.; Whitesides, G. M. *Langmuir* **1988**, *4*, 546.
8. Dialkyl thioethers were shown to form a ($\sqrt{3} \times \sqrt{3}$) R 30° lattice which is indistinguishable from n-alkane thiols. However, Strong and Whitesides (reference 7) attributed the observed structure to thiol impurities. More recently it was shown that highly purified dialkyl thioethers (e. g. H33C16-S C16H33) form SAMs that show the ($\sqrt{3} \times \sqrt{3}$) R 30° tail group lattice. In addition, thioethers were shown to corrode the gold similar to thiols. (Huisman, B.-H. Ph.D. Thesis, University of Twente, The Netherlands, 1998.; Schönherr, H.; Huisman, B.-H.; van Veggel, F. C. J. M.; Vancso, G. J.; Reinhoudt, D. N. *submitted*).
9. Liu, G.; Fenter, P.; Chidsey, C. E. D.; Ogletree, D. F.; Eisenberger, P.; Salmeron, M. *J. Chem. Phys.* **1994**, *101*, 4301.
10. Jaschke, M.; Schönherr, H.; Wolf, H.; Ringsdorf, H.; Besocke, M. K.; Bamberg, E.; Butt, H.-J. *J. Phys. Chem.* **1996**, *100*, 2290.
11. Nuzzo, R. G.; Zegarski, B. R.; Dubois, L. H. *J. Am. Chem. Soc.* **1987**, *109*, 733.
12. Sellers, H.; Ulman, A,; Shnidman, Y.; Eilers, J. E. *J. Am. Chem. Soc.* **1993**, *115*, 9389.
13. Evidence from He diffraction (Camillone III, N.; Chidsey, C. E. D.; Liu, G.-y.; Scoles, G. *J. Chem. Phys.* **1993**, *98*, 3503) as well as STM ((a) Poirier, G. E.; Tarlov, M. J. *Langmuir* **1994**, *10*, 2853; (b) Delamarche, E.; Michel, B.; Gerber, Ch.; Anselmetti, D.; Güntherodt, H.-J.; Wolf, H.; Ringsdorf, H. *Langmuir* **1994**, *10*, 2869) shows the existence of a c(4 × 2) superlattice. Recently it was shown that this superlattice is the lowest energy structure and that it can be induced by a STM or AFM tip (Touzov, I.; Gorman, C. B. *J. Phys. Chem. B* **1997**, *101*, 5263).

30

14. Wolf, H.; Ringsdorf, H.; Delamarche, E.; Takami, T.; Kang, H.; Michel, B.; Gerber, Ch.; Jaschke, M.; Butt, H. J.; Bamberg, E. *J. Phys. Chem.* **1995**, *99*, 7102.

15. Schönherr, H.; Kremer, F. J. B.; Kumar, S.; Rego, J. A.; Wolf, H.; Ringsdorf, H.; Jaschke, M.; Butt, H.-J.; Bamberg, E. *J. Am. Chem. Soc.* **1996**, *118*, 13051.

16. Chidsey, C. E. D.; Loiacono, D. N. *Langmuir* **1990**, *6*, 682.

17. In poly(tetraflurorethylene), the molecules form a 1*13/6 helix below room temperature (in the crystallographic direction along the polymer chain (c) there are 13 CF2 units, while the helix makes 6 turns corresponding to one crystallographic repeat unit).

18. Alves, C. A.; Porter, M. D. *Langmuir* **1993**, *9*, 3507 and references therein.

19. Nelles, G.; Schönherr, H.; Jaschke, M.; Wolf, H.; Schaub, M.; Küther, J.; Tremel, W.; Bamberg, E.; Ringsdorf, H.; Butt, H.-J. *Langmuir* **1998**, *14*, 808.

20. Schönherr, H.; Vancso, G. J. *Langmuir* **1997**, *13*, 3769.

21. Schönherr, H.; Ringsdorf, H. *Langmuir* **1996**, *12*, 3891.

22. Xiao, X.; Hu, J.; Charych, D. H.; Salmeron, M. *Langmuir* **1996**, *12*, 235.

23. Lio, A.; Charych, D. H.; Salmeron, M. *J. Phys. Chem. B* **1997**, *101*, 380.

24. Butt, H.-J.; Seifert, K.; Bamberg, E. *J. Phys. Chem.* **1993**, *97*, 7316.

25. Overney, R. M.; Takano, H.; Fujihira, M.; Paulus, W.; Ringsdorf, H. *Phys. Rev. Lett.* **1994**, *72*, 3546.

26. Liu, Y.; Wu, T.; Evans, D. F. *Langmuir* **1994**, *10*, 2241.

27. Carpick, R. W.; Salmeron, M. *Chem. Rev.* **1997**, *97*, 1163.

28. The atomic lattice of Au(111) on freshly annealed samples could readily be imaged. The lattice distances of 2.9 Å agree very well with the previouls published value for Au(111) (Manne, S.; Butt, H.-J.; Gould, C. A. C.; Hansma, P. K. Appl. Phys. Lett. 1990, 56, 1758.).

29. Schönenberger, C.; Sondag-Huethorst, J. A. M.; Jorritsma, J.; Fokkink, L. G. *Langmuir* **1994**, *10*, 611.

30. Miller, W. J.; Abbott, N. L. *Langmuir* **1998**

31. Al-Khairalla, B.; Evans, S. D. unpublished results.

32. Rabolt, J. F. private communication.

33. Lenk, T. J.; Hallmark, V. M.; Hoffman, C. L.; Rabolt, J. F.; Castner, D. G.; Erdelen, C.; Ringsdorf, H. *Langmuir* **1994**, *10*, 4610.

34. Israelachvili, J. N.; Chen, Y.-L.; Yoshizawa, H. in *Fundamentals of Adhesion and Interfaces*, Rimai, D. S.; DeMejo, L. P.; Mittal, K. L. (Eds.), VSP **1995**, pp. 261 - 279.

Chapter 3

Aggregation Structure and Surface Properties of Immobilized Fluoroalkylsilanes and Their Mixed Monolayers

Ken Kojio[1], Atsushi Takahara[1], and Tisato Kajiyama[2, *]

[1]Department of Applied Chemistry and [2]Institute for Fundamental Research of Organic Chemistry, Kyushu University, 6–10–1 Hakozaki, Higashi-ku, Fukuoka 812–8581, Japan

Stable fluoroalkyl, alkyl and mixed fluoroalkyl/alkyl monolayers were prepared on solid substrates by the Langmuir method. The n-octadecyltrichlorosilane (OTS), 18-nonadecenyltrichlorosilane (NTS) and [3-(perfluorododecyl)propyloxy]triethoxysilane (FDOPTES) monolayers were in a crystalline state at 293 K, as determined by electron diffraction (ED) studies. In contrast, the n-dodecyltrichlorosilane (DDTS) and [2-(perfluorooctyl)ethyl] -trichlorosilane (FOETS) monolayers with shorter hydrophobic chains were in an amorphous state at 293 K. Molecular images of the crystalline OTS and FDOPTES monolayers were acquired with high-resolution atomic force microscopy (AFM). The direct observation of the molecular arrangement of the crystalline OTS and FDOPTES monolayers was successfully achieved. The (10) spacings for these monolayers corresponded well to the cross section area of the alkyl and fluoroalkyl chains. It was revealed from the contact angle measurements that the magnitudes of surface free energies of the fluoroalkylsilane monolayers were higher than those of the alkylsilane monolayers. The surface free energy of the crystalline fluoroalkylsilane monolayer (FDOPTES monolayer) was higher than that of the amorphous monolayers due to high orientation of the fluoromethyl end groups at the monolayer surface. AFM observation revealed that the (crystalline OTS/amorphous FOETS) mixed monolayer was in a phase-separated state due to the crystallization of the OTS molecules. The (crystalline NTS/amorphous FOETS) mixed monolayer also showed a phase separation. By oxidation of the NTS phase in the (NTS/FOETS) mixed monolayer, a phase-separated structure with large surface energy gap can be constructed. The surface properties of the mixed monolayer was evaluated based on lateral force microscopy (LFM) and adhesion force measurement.

Introduction

Fluorinated monolayers have uses as diffusion barriers, low-energy contamination-resistant surfaces, bacterial plaque control, lubricants, etc. There have been a number of attempts to create ordered fluorinated monolayers through the use of the Langmuir method and self-assembly using fluoroalkane thiols.*(1-6)* A fluoroalkyl-grafted surface is expected to be highly hydrophobic due to the orientation of fluoromethyl group at the surface. When the conventional Langmuir monolayer is used for the preparation of fluoroalkyl-grafted surface from fluoroalkylcarboxylic acid, the surface structure is easily reorganized upon exposure to the water owing to the absence of strong interaction with substrate surface. Organosilanes have been applied as coupling agents because the organosilane compounds can be tightly immobilized onto the material surface through strong chemical interaction. When the alkyltrichlorosilane monolayer is polymerized on the water surface, the polyalkylsiloxane oligomer with a certain number of unreacted -OH groups was formed. This polyalkylsiloxane monolayer can be immobilized onto the silicon wafer substrate by upward drawing from the air/water interface.*(7-17)* This approach can be also applied to the preparation of the stable fluoroalkylsilane monolayers.

We have reported that the alkylsilane and fluoroalkylsilane mixed monolayer prepared by the Langmuir method formed a phase-separated island structure.*(7-9)* This mixed monolayer forms a phase-separated surface with different surface free energies. Also, since the reaction of the vinyl end group of the hydrophobic chain can be used to introduce various functional groups such as carboxyl and hydroxy groups,*(17-19)* surfaces with large surface free energy gaps can be constructed by utilizing the phase-separated surface of vinyl and fluoromethyl groups.

In this study, the molecular aggregation states of the alkyltrichlorosilane, fluoroalkylsilane and their mixed monolayers prepared by the Langmuir method were investigated on the basis of transmission electron microscopy (TEM) and atomic force microscopy (AFM). The surface properties of monolayers were characterized by contact angle measurements and lateral force microscopy (LFM). Finally, the preparation of a novel phase-separated structure with hydrophilic-domain and hydrophobic matrix is presented.

Experimental

Monolayer Preparation   n-Octadecyltrichlorosilane (OTS, $CH_3(CH_2)_{17}SiCl_3$: PCR, Ltd., Co.), n-dodecyltrichlorosilane (DDTS, $CH_3 (CH_2)_{11}SiCl_3$: PCR, Ltd., Co.),

18-nonadecyltrichlorosilane (NTS, $CH=CH_2(CH_2)_{17}SiCl_3$: Shin-Etsu Chemical, Ltd., Co.), [2-(perfluorooctyl)ethyl]trichlorosilane (FOETS, $CF_3(CF_2)_7(CH_2)_2SiCl_3$: Shin-Etsu Chemical, Ltd., Co.), and [3-(1H,1H,2H,2H-perfluorododecyl)propyloxy] triethoxysilane (FDOPTES, $CF_3(CF_2)_9(CH_2)_2O(CH_2)_3Si(OC_2H_5)_3$) were used to prepare the monolayer. FDOPTES was synthesized from 1H,1H,2H,2H-perfluorodecylalcohol, vinyl chloride and triethoxyhydrosilane.*(13)* The organosilane compounds were purified by vacuum distillation. A toluene solution of organosilane was prepared with a concentration of 3 x 10^{-3} M. Toluene was refluxed with sodium and distilled under atmospheric pressure in order to avoid a reaction of organosilanes with moisture. The organosilane toluene solutions were spread on the pure water surface at a subphase temperature of 293 K at pH of 5.8 (for trichlorosilane) and 1.0 (for triethoxysilane). Surface pressure-area (π-A) isotherms were measured with a computer-controlled home made Langmuir-trough. To form the polymeric monolayer, the spread molecules were kept on the water subphase under a certain constant surface pressure for 15 minutes. After forming the polymeric monolayer on the water subphase, it was transferred onto substrates by the upward drawing method, and strongly immobilized onto the substrate with Si-OH groups.

Oxidation of NTS Monolayer Oxidation of the vinyl end group of the NTS monolayer was carried out to prepare the monolayer surface with carboxyl groups.*(17,19)* Stock solutions of $KMnO_4$ (5 mM), $NaIO_4$ (195 mM) and K_2CO_3 (18 mM) in water were prepared and then, the oxidizing solution was obtained by mixing 1 mL from each of these solutions and 7 mL of distilled water. The NTS monolayer was immersed in this oxidizing solution for a few minutes. Then, the samples were removed from the solution and sequentially rinsed in 20 mL of $NaHSO_3$ (0.3 M), distilled water, 0.1 N HCl and ethanol.

Electron Diffraction (ED) To study the molecular aggregation state of the monolayer with TEM, the monolayer must be transferred onto the substrate without any change of the aggregation or crystallographic structure of the monolayer on the water subphase.*(20)* Therefore, the monolayer was transferred by the upward drawing method onto the collodion-covered electron microscope grids (200 mesh) covered with an evaporated hydrophilic SiO layer. The ED pattern was obtained with TEM (Hitachi H-7000) operated at an acceleration voltage of 75 kV, a beam current of 0.5 μA and an electron beam spot size of 10 μm diameter.

Infrared spectroscopic (IR) measurement IR measurement was performed to evaluate the conformation of the alkyl chain of the organosilane monolayer in the transmission mode with an unpolarized beam. The organosilane monolayers were transferred on a wedged silicon wafer to eliminate the multiple reflection of the infrared beam within the silicon wafer substrate.*(21)* The spectra were recorded at a resolution of 4 cm^{-1} at 293 K with a Nicolet Magna 860 instrument which was equipped with a mercury-cadmium-telluride (MCT) detector. To obtain the spectra with high signal-to-noise ratios, 256 scans were collected.

Scanning Force Microscopic (SFM) Observation The high-resolution AFM images of the monolayers were observed with AFM (SPA 300, Seiko Instruments

Industry, Co., Ltd., Japan) in water at room temperature, using a 1 μm x 1 μm scanner and a silicon nitride tip attached to a cantilever with a spring constant of 0.025 N m[-1]. In water, the effect of adhesion force between tip and sample surface due to adsorbed water can be eliminated in AFM imaging. AFM was operated in the constant height mode. Also, AFM observation was performed for the monolayers transferred onto a silicon wafer at various surface pressures. The AFM was operated under constant force mode under air at room temperature, using a 20 μm x 20 μm scanner and a silicon nitride tip on a cantilever with a spring constant of 0.025 N m[-1]. LFM measurement was carried out to investigate the surface mechanical properties of the organosilane monolayers using a silicon nitride tip on a cantilever with a spring constant of 0.09 N m[-1]. The cantilever was scanned perpendicular to the axis of cantilever to enlarge the torsion due to interaction between tip and sample surface. The mapping of the adhesion force was performed to evaluate the surface properties by measuring the force-distance curve at each point. AFM adhesion force maps were obtained at a resolution of 100 x 100 x 100 points (x-y-z directions) with Topometrix Explorer.

Contact angle measurement   To evaluate the surface free energy of the organosilane monolayers, static contact angle measurements were carried out. Contact angles of water and methylene iodide droplets on the monolayer surface were measured with a contact angle goniometer CA-D (Kyowa-Kaimenkagaku, Co., Ltd.) at 293 K.

The dynamic contact angle (DCA) of the monolayers against water was measured by a dynamic Wilhelmy plate technique.*(22-24)*   Water for DCA measurement was purified with a Milli-Q system. The surface tension was measured with high-sensitivity strain gauge. The monolayers were immobilized onto a double polished silicon wafer substrate (20 x 40 mm²). The contact angle hysteresis curve was obtained at 293 K under a testing speed of 10 mm min[-1] by plotting the observed force against the immersion depth.

Results and discussion

Molecular Aggregation State Based on Electron Diffraction   Figure 1 shows the π-A isotherms for the DDTS, OTS, NTS, FOETS and FDOPTES monolayers on the pure water surface at 293 K. The π-A isotherms for the OTS, NTS and FDOPTES monolayers showed a steep increase in surface pressure with a decrease in surface area. The molecular occupied areas, that is, the limiting areas were determined to be 0.24 nm² molecule[-1], 0.23 nm² molecule[-1] and 0.30 nm² molecule[-1], respectively, for the OTS, NTS and FDOPTES monolayers. On the other hand, the π-A isotherms of the DDTS and FOETS monolayers showed a gradual increase in surface pressure with a decrease in a surface area. The molecular occupied areas of the DDTS and FOETS monolayers were slightly larger than those of the OTS and FDOPTES monolayers, respectively.

The molecular aggregation states of the monolayers were investigated on the basis of ED patterns. Figure 1 also shows the ED patterns of the DDTS, OTS, NTS, FOETS and FDOPTES monolayers which were transferred onto a hydrophilic SiO substrate at various surface pressures at 293 K.*(7,16,17)*   The ED patterns of the

OTS and FDOPTES monolayers exhibited hexagonal crystalline arcs which were broadened along the azimuthal direction.   These ED patterns indicated that the OTS and FDOTPES molecules formed crystalline monolayers with slight sintering behavior at the domain interface during compression.   Also, the ED pattern of the NTS monolayer exhibited hexagonal spots, indicating the formation of large crystalline monodomains in the NTS monolayer at 293 K.   That is, this result indicates that fairly large crystalline NTS domains, in comparison with an electron beam diameter of 10 μm, were formed by a surface pressure-induced sintering mechanism such as fusion or recrystallization at the domain interface during compression of the monolayer on the water subphase.   The ED patterns of the OTS and NTS monolayers showed the (10) spacing of ca. 0.42 nm.   The (10) spacings of the OTS and NTS monolayers agree with that of the stearic acid (SA) monolayer.*(25,26)*   Also, the (10) spacing of the crystalline FDOPTES monolayer estimated from ED pattern to be ca. 0.50 nm.   This spacing corresponds to that observed for a fluoroalkyl thiol monolayer with high-resolution AFM observation.*(27)*   On the other hand, the ED patterns of the DDTS and FOETS monolayers showed amorphous halos at 293 K.   These results corresponded to those obtained by π-A isotherm measurements.

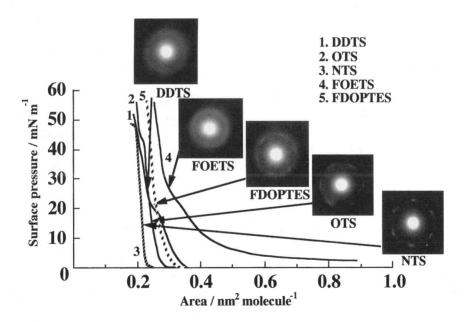

Figure 1. Surface pressure-area (π-A) isotherm and electron diffraction (ED) patterns for the DDTS, OTS, NTS, FOETS, and FDOPTES monolayers on the water subphase at the temperature of 293 K. The ED patterns of the monolayer were taken after transfer onto a hydrophilic SiO substrate around surface pressures of 20 mN m^{-1}.

Molecular Aggregation State Based on High-resolution AFM Observation
High-resolution AFM observation was performed to evaluate directly the molecular arrangement for the crystalline OTS and FDOPTES monolayers. Figures 2 (a) and (b) show the low pass-filtered high-resolution AFM images for the crystalline OTS and FDOPTES monolayers transferred onto a silicon wafer substrate by the upward drawing method at a surface pressure of 15 mN m^{-1} at 293 K.(11,16) The (10) spacings for OTS and FOETS monolayers were estimated to be ca. 0.42 nm and 0.50 nm on the basis of the two-dimensional fast Fourier transform (2D-FFT), respectively. These magnitudes well agreed with the (10) spacing determined from the ED patterns for the crystalline OTS and FDOPTES monolayers shown in Figure 1. Therefore, the agreement of the (10) spacing from the ED patterns and the high-resolution AFM images indicate that the higher portions (the bright dots) in the AFM images of Figures 2 (a) and (b) represent the individual methyl and fluoromethyl groups of the OTS and FDOPTES molecule in the monolayers.

Surface Properties of Organosilane Monolayers The surface free energies of the organosilane monolayers were evaluated on the basis of a static contact angle measurement for water and methylene iodide. Furthermore, the DCA of the organosilane monolayer was measured with the Wilhelmy plate technique.(21-23) The DCA observed on advancement of the liquid front is called the advancing contact angle, θ_a, and the angle of the receding liquid front is the receding contact angle, θ_r. The DCA data reflect the aspects of surface molecular mobility, surface roughness, and surface heterogeneity at the air-water-solid interface. Table 1 and Figure 3 show the contact angles for water and methylene iodide, the surface free energies by static contact angle measurement and the advancing and receding contact angles for water by DCA measurement of the organosilane monolayers. The surface free energies of the fluoroalkylsilane monolayers were smaller than those of the alkylsilane monolayers. This result corresponds well to the relationship between polyethylene (PE) and polytetrafluoroethylene (PTFE). The surface free energies of the crystalline OTS, NTS and FDOPTES monolayers were smaller than those of the amorphous DDTS and FOETS monolayers, respectively. Since molecular ordering in the amorphous monolayer is low, the hydrophobic end groups were not completely oriented at the monolayer surface. Furthermore, the crystalline OTS and FDOPTES monolayers showed smaller surface free energies than the polymers which have similar chemical structure, that is, PE and PTFE. These results indicate that the hydrophobic end groups of the crystalline OTS and FDOPTES monolayers are located at the monolayer surface. The magnitude of surface free enrgy of the crystalline NTS monolayer with hydrophobic vinyl end group was 25.0 mN m^{-1}. The magnitude of the surface free energy of the NTS$_{COOH}$ monolayer was 68.8 mN m^{-1}, indicating the oxidation from the vinyl end group to carboxyl group. The oxidation of the vinyl end group was confirmed by X-ray photoelectron spectroscopic (XPS) measurements.(17) The advancing contact angles of the FDOPTES and FOETS monolayers with fluoroalkyl chain were larger than those of the OTS and DDTS monolayers with alkyl chain. On the other hand, receding contact angles of the FDOPTES and FOETS monolayers were smaller than those of

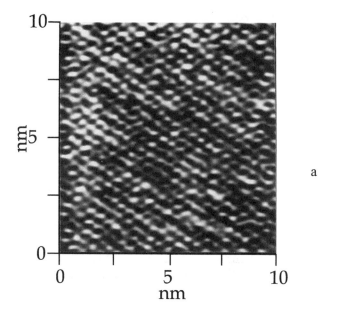

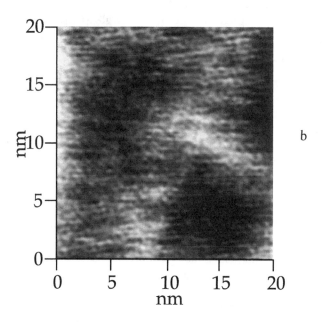

Figure 2. The high resolution low pass filtered AFM images for (a) the crystalline OTS and (b) FDOPTES monolayers over a scan area of 10 x 10 nm^2. The monolayer was prepared on a silicon wafer substrate by the upward drawing method at a surface pressure of 15 mN m^{-1} at the subphase temperature of 293 K. Note the periodic arrangement of the molecules with a hexagonal array.

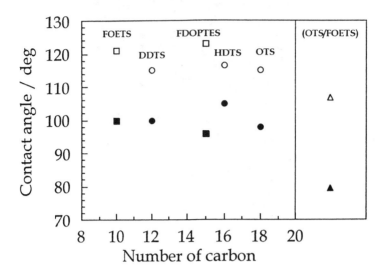

Figure 3. Dynamic contact angles (advancing and receding contact angles) for water of the organosilane monolayers. Open and solid symbols represent advancing and receding contact angles, respectively. Circles, squares and triangles symbols represent contact angles for the alkylsilane, fluoroalkylsilane, (OTS/FOETS)(50/50)mixed monolayers, respectively.

the OTS and DDTS monolayers.   It is possible that the dipole of the fluoromethyl end groups of the fluoroalkylsilane interact with that of the water.   The hydrogen bond component of the surface free energy of the fluoroalkylsilane monolayer is larger than that of the alkylsilane one as shown in Table 1.   The hysteresis of the advancing and receding contact angles of the NTS monolayer was smallest of all the organosilane monolayers.   This result corresponds well to result that the coverage of the crystalline NTS monolayer is above 99 % and homogeneous in comparison with other organosilane monolayers.(17)   The advancing and receding contact angles of the (OTS/FOETS) mixed monolayer were smaller than the organosilane homomonolayers.   Also, the hysteresis of the (OTS/FOETS) mixed monolayer was the largest.   This result indicates that the distribution of phases with different surface energy affects the magnitude of the total surface free energies of the monolayer.

(OTS/FOETS) Mixed Monolayer   The mixed monolayers were prepared as a 2-dimensional analogue of phase separation in a polymer blend.   It is well known that mixture of alkane and fluoroalkane induces a macroscopic phase separation. However, due to the strong lateral interaction among hydrophilic silanol groups, the macroscopic phase separation cannot occur.   Figure 4 shows the AFM images for the (OTS/FOETS) (a) (25/75), (b) (50/50), and (c) (75/25) mixed monolayer transferred onto silicon wafer substrates by the Langmuir method at a surface

Table 1   The contact angle for water and methylene iodide and surface free energies of the organosilane monolayers by static contact angle measurement

	θ_{H_2O}/deg	$\theta_{CH_2I_2}$ /deg	Surface free energy/mN m^{-1}		
			γ_s^h	γ_s^d	γ_s
OTS	109	76	0.3	19.5	19.8
DDTS	108	74	0.3	20.3	20.6
FDOPTES	112	100	2.0	7.5	9.5
FOETS	110	96	1.8	9.0	10.8
NTS	95	63	1.6	25.4	26.9
NTS$_{COOH}$	24	25	35.6	33.7	68.8
(OTS/FOETS)	96	86	5.2	12.2	17.4

pressure of 25 mN m^{-1} at 293 K and (d) the height profile along the line shown in (b). It was revealed that the (OTS/FOETS) mixed monolayers were in a phase-separated state, and circular flat-topped domains of ca. 1-2 μm in a diameter were surrounded by a sea-like and flat region.   Since the area occupied by the circular flat-topped domains increases with an increase in the OTS content, it is expected that the circular flat-topped domain corresponds to the OTS domain.   It was revealed from Figure 4(d) that the circular flat-topped domains were 1.1-1.3 nm higher than the surrounding flat monolayer.   Since the difference in molecular lengths between OTS and FOETS is ca. 1.3 nm, it can be concluded that the higher, circular domains and the surrounding flat matrix regions were composed of OTS and FOETS molecules, respectively.   OTS molecules formed circular domains even if the molar percent of OTS molecules was 75 % as shown in Figure 4(c).   It is apparent from the ED pattern of the (OTS/FOETS) (75/25) mixed monolayer that the OTS domain is in a crystalline state, since the magnitude of spacing corresponds to the (10) spacing of the OTS monolayer.   Transmission infrared (IR) measurements were done to evaluate the conformation of the alkyl chain of the organosilane monolayer. The organosilane monolayers were transferred onto wedged silicon wafer.   Since fringing effects in IR spectra can be minimized by use of wedged silicon wafer.(21) IR measurement were performed in the transmission mode with an unpolarized beam. The peaks assigned to the antisymmetric ($v_a(CH_2)$) and the symmetric ($v_s(CH_2)$) stretching bands of alkyl chains of the OTS molecule were observed at 2917 and 2849 cm^{-1}, in the IR spectra of (OTS/FOETS) mixed monolayer, respectively.

40

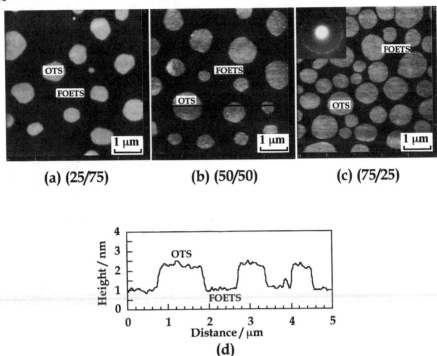

(a) (25/75)　　　　(b) (50/50)　　　　(c) (75/25)

(d)

Figure 4. AFM images for the (OTS/FOETS) (a) (25/75), (b) (50/50), (c) (75/25) mixed monolayer prepared onto silicon wafer substrates by the Langmuir method at a surface pressure of 25 mN m^{-1} at 293 K. (d) The height profile along the line shown in (b).

These wavenumbers indicate that alkyl chains of the OTS molecule in the mixed monolayer are in an all-trans state and are packed in crystal lattice.[28]　The peaks of the $\nu_a(CH_2)$ and the $\nu_s(CH_2)$ bands of the OTS molecules in the (OTS/FOETS) mixed monolayer were also observed at the same wavenumbers.　Therefore, it is considered that the OTS molecules in the (OTS/FOETS) mixed monolayer are in an all-trans state and form crystalline domains.　Several mixed monolayers were prepared to elucidate the reason for the phase separation.　AFM observation revealed that the (OTS(crystalline alkylsilane)/DDTS(amorphous alkylsilane)) mixed monolayer showed phase separation.[9]　In contrast, the (DDTS(amorphous alkylsilane)/FOETS(amorphous fluoroalkylsilane)) mixed monolayer did not show phase separation.[8,9]　Therefore, it is reasonable to conclude that the phase separation in the (OTS/FOETS) mixed monolayer might be due to the crystallizability of the OTS molecules.

(NTS/FOETS) Mixed Monolayer　　The (crystalline organosilane/amorphous organosilane) mixed monolayers showed a phase-separated structure on the water subphase as mentioned above. Therefore, it is also expected that the (crystalline NTS/amorphous FOETS) mixed monolayer would form a phase-separated structure. Figure 5 (a) shows the AFM image of the (NTS/FOETS) mixed monolayer prepared on the water subphase at 293 K and transferred onto a silicon wafer substrate at the surface pressure of 20 mN m^{-1}. The (NTS/FOETS) mixed monolayer showed that circular flat-topped domains of ca. 1-3 μm in a diameter were surrounded by a sea-like matrix, similar to the (OTS/FOETS) monolayer.[7-9] The number and the area fraction of the domain increased with an increase in the NTS molar fraction. Figure 5 (c) shows the height profile along the line shown in Figure 5 (a). The height difference between domain and matrix phases was ca. 1.2-1.4 nm, which corresponded to the difference in their molecular lengths. Therefore, it seems reasonable to conclude that the domain and the matrix phases in the (NTS/FOETS) mixed monolayer are composed of the NTS and the FOETS molecules, respectively. Furthermore, LFM measurement was carried out to investigate the surface mechanical properties of the mixed monolayer surface. Figure 5 (b) shows the LFM image of the (NTS/FOETS) mixed monolayer. The brighter and darker parts correspond to the regions with higher and lower lateral force. Then Figure 5 (b) shows that the magnitude of the lateral force of FOETS phase was higher than that of the NTS domain. In the case of the (OTS/FOETS) mixed monolayer, it was reasonably concluded that the FOETS matrix phase showed higher lateral force than the OTS domains due to higher shear strength between sample surface and cantilever tip, that might originate from the rigid rod-like conformation of the fluoroalkyl group.[9,29] Therefore, it seems reasonable to consider that the lateral force of the FOETS matrix phase was higher than that of the NTS domain due to a similar reason.

(NTS$_{COOH}$/FOETS) Mixed Monolayer Due to the strong anchoring between the silanol groups and Si-wafer surface, the monolayer shows highly stability in various solvents. Highly stable hydrophobic-hydrophilic phase-separated monolayer was prepared by the oxidation of the NTS phase in the (NTS/FOETS) mixed monolayer. Figures 6 (a) and (b) show the AFM and the LFM images of the (NTS$_{COOH}$/FOETS) mixed monolayer prepared by oxidizing the NTS phase in the (NTS/FOETS) mixed monolayer surface. Figure 6 (c) shows the height profile along the line shown in Figure 6 (a). It is apparent that the surface morphology of the (NTS/FOETS) mixed monolayer was not changed after oxidation. The height difference between NTS$_{COOH}$ domain and FOETS matrix phases in the (NTS$_{COOH}$/FOETS) mixed monolayer was the same as that for the (NTS/FOETS) mixed monolayer. Also, XPS measurements performed on the (NTS/FOETS) and the (NTS$_{COOH}$/FOETS) mixed monolayers confirmed the oxidation of the NTS phase. Since the ratio of oxygen/carbon for the (NTS$_{COOH}$/FOETS) mixed monolayer was larger than that for the (NTS/FOETS) one, it was confirmed that the vinyl end groups of the NTS molecules were oxidized to carboxyl groups. Also, the contact angle against water of the NTS monolayer was decreased by oxidation. The lateral force of the NTS$_{COOH}$ phase was higher than that of the FOETS phase in the case of the (NTS$_{COOH}$/FOETS) mixed monolayer as shown in Figure 6 (b), in contrast to the case

42

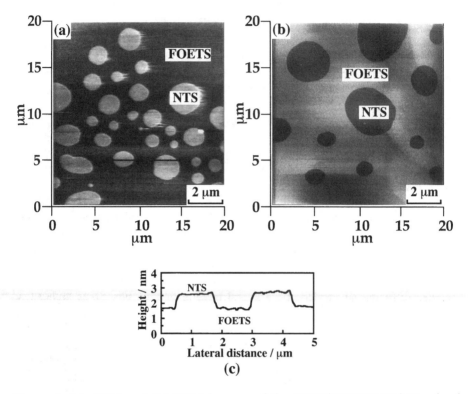

Figure 5. (a) AFM and (b) LFM images of the (NTS/FOETS)(50/50) mixed monolayer prepared on the water subphase at 293 K and transferred onto silicon wafer substrate at the surface pressure of 20 mN m^{-1}. (c) Height profile along the line shown in (a).

of the (NTS/FOETS) mixed monolayer.   As the NTS$_{COOH}$ phase had hydrophilic carboxyl end groups at the surface, it is reasonable to consider that these end groups can form intermolecular hydrogen bonds among neighboring COOH groups of the NTS$_{COOH}$ molecules.   Therefore, the surface of the outermost NTS$_{COOH}$ phase is expected to show a higher shear strength than for the NTS phase.   Also, the magnitude of the lateral force corresponds to the sum of frictional force and adhesion force between the sample and sliding cantilever tip.   Guckenberger et al. reported that the enhanced adhesion force due to water capillary force acted as a predominant component of the normal force interacting between the material surface and cantilever tip.*(30)*   Also, Fujihira et al. reported that the adhesion force strongly affected to the lateral force in the case of a Si substrate covered partially with OTS.*(31)*   Since the adsorbed water layer on the hydrophilic NTS$_{COOH}$ surface might be thicker than that of the hydrophobic surface, the water capillary force interacting between NTS$_{COOH}$ monolayer surface and hydrophilic Si$_3$N$_4$ tip strongly

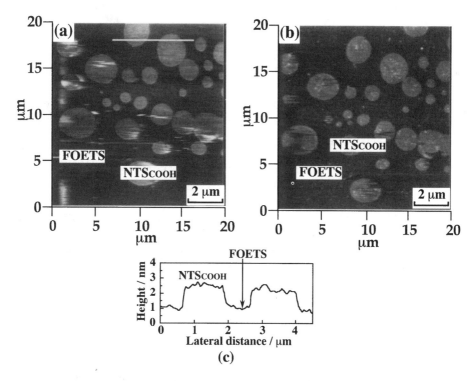

Figure 6. (a) AFM and (b) LFM images of the (NTS$_{COOH}$/FOETS) (50/50) mixed monolayer prepared by oxidizing NTS phase in the (NTS/FOETS) (50/50) mixed monolayer surface. (c) Height profile along the line shown in (a).

contributed to the adhesion force of the NTS$_{COOH}$ phase. Therefore, it is reasonable to consider that the NTS$_{COOH}$ phase exhibited higher lateral force than the FOETS one due to the formation of intermolecular hydrogen bonding and thicker absorbed water layer as discussed above. As the result, the phase-separated monolayer with large surface energy gap of 55 mN m^{-1} could be constructed.

To map the adhesion of the (NTS/FOETS) and (NTS$_{COOH}$/FOETS) mixed monolayers, the force-distance curve was measured at a resolution of 100 x 100 (x-y direction) data points and 100 (z direction) data points for each force-distance curve. Figure 7 shows the adhesion force images of (a) the (NTS/FOETS) and (b) the (NTS$_{COOH}$/FOETS) (b) mixed monolayers. The brighter and darker phases correspond to lower and higher adhesion force regions, respectively. Although both the NTS and FOETS phases in the (NTS/FOETS) mixed monolayer are hydrophobic, the difference in adhesion force was clearly observed as shown in Figure 7 (a). A similar result was obtained for the (OTS/FOETS) mixed monolayer. Since based on contact angle measurements the surface free energies of the NTS and the FOETS monolayers were 26.9 mN m^{-1} and 11.2 mN m^{-1}, it is likely that the thickness of the adsorbed water layer on the NTS monolayer surface is slightly thicker than the FOETS monolayer surface. Therefore, the adhesion interaction between the cantilever tip and the NTS phase was larger than for the FOETS phase. The adhesion force image of the (NTS$_{COOH}$/FOETS) mixed monolayer also showed a similar contrast as shown in Figure 7 (a). The NTS$_{COOH}$ domain showed higher adhesion force than the FOETS phase. It is considered that water capillary force of the NTS$_{COOH}$ phase is larger because of its hydrophilicity. Figure 8 shows the representative force-distance curves for (a) the (NTS/FOETS) and (b) the (NTS$_{COOH}$/FOETS) mixed monolayers at each phase. The adhesion force of the NTS$_{COOH}$ phase is clearly larger than those of the NTS and the FOETS phases. The average (200 points) of the output current of photodiode corresponding to the cantilever deflection was normalized by FOETS phase. As the result, the ratio of the average of the output current of the FOETS, NTS and NTS$_{COOH}$ phases were 1, 1.2 and 1.71, respectively. Since the surface free energy of the NTS$_{COOH}$ monolayer was 68.8 mN m^{-1}, it is inferred that the water adsorbed on the monolayer surface is very thick in comparison with those of the hydrophobic NTS and FOETS monolayer surfaces. Therefore, it seems reasonable to conclude that the adhesion force of the NTS$_{COOH}$ phase is very large due to the water capillary force interacting between the cantilever and monolayer surface. This result corresponds well to that of the friction force measurements (Figure 6).

Conclusion

Stable monolayers of fluoroalkylsilane, alkylsilane and their mixtures on the solid substrate were prepared by the Langmuir method. The OTS, NTS and FDOPTES monolayers were in a crystalline state at 293 K. The DDTS and FOETS monolayers with shorter hydrophobic chains were in an amorphous state at 293 K. Molecular images of the crystalline OTS and FDOPTES monolayers were successfully observed. The magnitude of the surface free energy of the crystalline FDOPTES monolayer was smallest in the organosilane monolayers. The (OTS/FOETS) mixed monolayer showed a phase-separated structure due to the crystallizability of the OTS molecules. The (NTS/FOETS) mixed monolayer also showed phase separation. By oxidation of the NTS phase in the (NTS/FOETS) mixed monolayer, a stable phase-separated structure with a large surface energy gap was successfully prepared.

Acknowledgment

This study was partially supported by a Research Fellowship of the Japan Society for the Promotion of Science for Young Scientists, and by a Grant-in-Aid for COE Research and Scientific Research on Priority Areas, "Electrochemistry of Ordered Interfaces" (No. 282/09237252), from Ministry of Education, Science, Sports and Culture of Japan. The FOETS was kindly supplied by Shin-Etsu Chemical Ltd., Co..

Literatures cited

1. Ulman, A. *An Introduction to Ultrathin Organic Films : Form Langmuir-Blodgett to Self-Assembly* (Academic, New York, 1991).

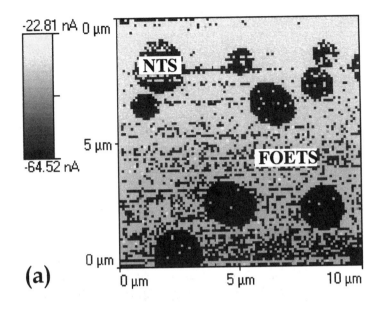

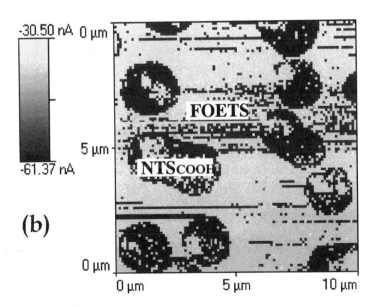

Figure 7. Adhesion force images of (a) the (NTS/FOETS) (50/50) and (b) the (NTS$_{COOH}$/FOETS) (50/50) (b) mixed monolayers.

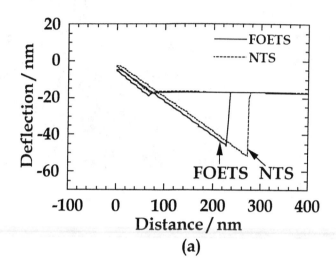

(a)

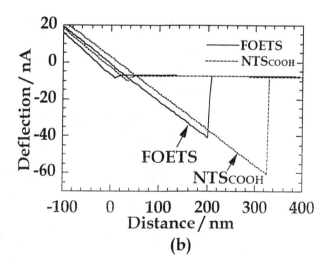

(b)

Figure 8. Force-distance curves for each phase in (a) the (NTS/FOETS) (50/50) and(b) the (NTS$_{COOH}$/FOETS) (50/50) mixed monolayers.

47

2. Yoshino, N.; Teranaka, T. *J. Biomater. Sci. Polym. Ed.* **1997**, *8*, 623.
3. Kim, H. I.; Koini, T.; Lee, R. T.; Perry, S. S. *Langumir* **1997**, *13*, 7192.
4. Hoffmann, W. P.; Stelzle, M.; Rabolt, F. J. *Langmuir* **1997**, *13*, 1877.
5. Lenk, T. J.; Hallmark, V. M.; Hoffmann, C. L.; Rabolt, J. F.; Castner, D. G.; Erdelen, C.; Ringsdorf, H. *Langmuir* **1994**, *10*, 4610.
6. Wang, W.; Castner, D. G.; Grainger, D. W. *Supramol. Sci.* **1997**, *4*, 85.
7. Ge, S. R.; Takahara, A.; Kajiyama, T. *J. Vac. Sci. Technol.* **1994**, *A 12*, 2530.
8. Ge, S. R.; Takahara, A.; Kajiyama, T. *Langmuir* **1995**, *11*, 1341.
9. Kajiyama, T.; Ge, S. R.; Kojio, K.; Takahara, A. *Supramol. Sci.* **1996**, *3*, 123.
10. Takahara, A.; Kojio, K.; Ge, S. R.; Kajiyama, T. *J. Vac. Sci. Technol.* **1996**, *A14*, 1747.
11. Kojio, K.; Ge, S. R.; Takahara, A.; Kajiyama, T. *Langmuir* **1998**, *14*, 971.
12. Takahara, A.; Ge, S. R.; Kojio, K.; Kajiyama, T. *Scanning Probe Microscopy of Polymers* ; ACS Symp. Ser. 694; Ratner, B. D.; Tsukruk, V. V. Eds.; American Chemical Society: Washington, DC, **1998**; Chapter 12.
13. Kojio, K.; Ge, S. R.; Takahara, A.; Kajiyama, T. *Rept. Prog. Polym. Phys. Jpn.* **1995**, *38*, 371.
14. Kojio, K.; Takahara, A.; Omote, K.; Kajiyama, T. *Langmuir* **2000**, *16*, 3932.
15. Kojio, K.; Takahara, A.; Kajiyama, T. *Langmuir* **2000**, *16*.
16. Kojio, K.; Takahara, A.; Kajiyama, T. *Coll & Sur. A* **2000**, *180*, 294.
17. Kojio, K.; Takahara, A.; Kajiyama, T. *Silicones and Silicon-Modified Materials*; Clarson, S. J., Ed.; American Chemical Society: Washington, DC, **2000**; Chapter 22.
18. Maoz, R.; Sagiv, J.; Degenhardt, D.; Möwald, H.; Quint, P. *Supramol. Sci.* **1995**, *2*, 9.
19. Wasserman, S. R.; Tao, Y.-T.; Whitesides, G. M. *Langmuir* **1989**, *5*, 1074.
20. Kajiyama, T.; Oishi, Y.; Uchida, M.; Morotomi, N.; Ishikawa, J.; Tanimoto, Y. *Bull. Chem. Soc. Jpn.* **1992**, *65*, 864.
21. Allara, D. L.; Parikh, A. N.; Rondelez, R. *Langmuir* **1995**, *11*, 2357.
22. Smith, L.; Doyle, C.; Gregonis, D. E.; Andrade, J. D. *J. Appl. Polym. Sci.* **1982**, *26*, 1269.
23. Takahara, A.; Jo, N. J.; Kajiyama, T. *J. Biomater. Sci. Polym. Edn.* **1989**, *1*, 17.
24. Kajiyama, T.; Teraya, T.; Takahara, A. *Polym. Bull.* **1990**, *24*, 333.
25. Kajiyama, T.; Oishi, Y.; Uchida, M.; Morotomi, N.; Kozuru, H. *Langmuir* **1992**, *8*, 1563.
26. Kajiyama, T.; Oishi, Y.; Uchida, M.; Takashima, Y. *Langmuir* **1993**, *9*, 1978.
27. Alves, C. A.; Porter, M. D. *Langmuir* **1993**, *9*, 3507.
28. MacPhail, R. A.; Strauss, H. L.; Snyder, R. G; Elliger, C. A; *J. Phys. Chem.* **1984**, *88*, 334.
29. Briscoe, J. B.; Evans, B. D. C. *Proc. R. Soc. Lond.* **1982**, *A 380*, 389.
30. Guckenberger, R.; Manfred, H.; Cevc, G.; Knapp, F. H.; Wiegrae, W.; Hillebrand, A. *Science* **1994**, *266*, 1538.
31. Fujihira, M.; Aoki, D.; Okabe, Y.; Takano, H.; Hokari, H.; Frommer, J.; Nagatani, Y.; Sakai, F. *Chem. Lett.* **1996**, 499.

Chapter 4

Stabilizing Colloids with Fluorocarbon–Hydrocarbon Diblocks: Example of Vesicles Made from Single-Chain Fluorinated Surfactants

Marie Pierre Krafft[1] and Françoise Giulieri[2]

[1]Chimie des Systèmes Associatifs, Institut Charles Sadron (CNRS), 6, rue Boussingault, 67 083 Strasbourg, cedex, France
[2]SOFFT, Unité de Chimie Moléculaire (associated to CNRS), University of Nice-Sophia Antipolis, 06 108 Nice, cedex 2, France

Small unilamellar vesicles made from a single-chain fluorinated phosphocholine, $C_8F_{17}(CH_2)_2OP(O_2)^-O(CH_2)_2N^+(CH_3)_3$ (F8H2PC) are very stable in pure water. They are, however, rapidly destroyed in the presence of a Hepes/NaCl buffer. Co-dispersion of the semi-fluorinated alkane $C_8F_{17}C_2H_5$ (F8H2) with F8H2PC (1:1 molar ratio) led to a homogenous population of vesicles (140 ± 37 nm) that were stable in the Hepes/NaCl buffer. No significant variation of average vesicle diameter and size distribution was observed after 3 months at 25°C in the buffer. The membrane permeability of the F8H2PC/F8H2 vesicles was investigated by measuring the release of 5,6-carboxyfluorescein (CF) entrapped in the internal space. It was found that 80% of the CF was still present in the reinforced fluorinated vesicles after 52 ± 6 min at 25°C, whereas immediate release (< 2 min) of the dye was observed when F8H2PC was alone. These results indicate that the F8H2 diblock is incorporated in the F8H2PC bilayer membrane in an orderly manner, and that it may prevent the dehydration of the polar head and the crystallization of the hydrophobic chain that usually result from the addition of salts on bilayer membranes made of single-chain surfactants. The combination of F8H2PC and F8H2 probably leads to the formation of a pseudo double-tailed entity, which is stabilized in the bilayer by strong hydrophobic and lipophobic interactions between the fluorinated segments of the two membrane components.

One of our on-going objectives is to investigate the potential of fluorocarbon/hydrocarbon diblock molecules such as $C_nF_{2n+1}C_mH_{2m+1}$ (FnHm) as components of colloidal systems (*1*). These compounds are amphiphiles, not in the usual sense that refers to molecules which associate hydrophilic and lipophilic moieties, but in an extended sense where lipophilic and fluorophilic moieties are associated within the same molecular structure in the absence of hydrophilic moieties.

Such compounds have first been synthesized by Brace (*2*). Their aggregation behavior in both fluorocarbons and hydrocarbons has been explored. Turberg and Brady have reported that these diblocks can form micelles when dispersed in either fluorinated or hydrogenated solvents, depending on the length of the Fn and Hm segments (*3*). Various investigations have been carried out on the structure and properties of FnHm diblocks in the solid state (*4-6*). Twieg *et al* have shown that FnHm diblock solutions (in either fluorinated or hydrogenated solvents), when cooled down below a certain transition temperature, can form gels constituted of a fibrous network entrapping large amounts of solvents (*7*).

More recently FnHm diblocks have been utilized as components of colloidal systems in several occasions. They were shown to provide a highly effective means of stabilizing phospholipid-based fluorocarbon-in-water emulsions (*8*), as well as a means of precisely controlling the emulsion's particle size (*9*). Such emulsions have potential as injectable oxygen delivery systems (*10,11*). FnHm diblocks allow the preparation of stable apolar hydrocarbon-in-fluorocarbon emulsions with potential for the delivery of lipophilic drugs through the pulmonary route (*12*). Microemulsions of $C_nF_{2n+1}CH_2CH=CHC_mH_{2m+1}$ using a commercial hydrogenated surfactant have also been reported (*13*). When incorporated into liposomes made from standard phospholipids FnHm diblocks strongly increased the stability of these liposomes: SUVs made from dimyristoyl phosphatidylcholine (DMPC) / FnHm mixtures become heat-sterilizable (*14*). FnHm diblocks also decrease the permeability of dipalmitoyl phosphatidylcholine (DPPC) vesicles with respect to the release of encapsulated 5,6-carboxyfluorescein (*15*). Incorporation of FnHm diblocks in the membrane of dipalmitoyl phosphatidylserine (DPPS) SUVs was found to reduce the rate of fusion of these SUVs, when induced by calcium ions (*16*). For appropriate values of n and m, the presence of FnHm diblocks in liposomes made from DMPC or DPPC was found to inhibit the enzymatic hydrolysis of the phospholipid by pancreatic porcine phospholipase A_2 (*17*).

Another project led to the finding that some single-chain fluorinated amphiphiles were able to form vesicles without the need of any additive or recourse to complementary associative interactions (hydrogen bonding, electrostatic forces, polymerization, etc.) destined to promote the alignment and packing of the molecules in the bilayer (*18*). Such surfactants include the perfluoroalkylated phosphocholine derivatives $C_nF_{2n+1}(CH_2)_mOP(O_2)^-O(CH_2)_2N^+(CH_3)_3$ (FnHmPC). For example, F8H2PC, when dispersed in water, readily forms SUVs (it only requires a few minutes of sonication). On the other hand, its hydrocarbon analogs $C_mH_{2m+1}OP(O_2)^-O(CH_2)_2N^+(CH_3)_3$ (HmPC), with m = 10 or 15, respectively, form micelles or crystallize when submitted to the same conditions (*18*). The F8H2PC vesicles are stable enough to resist heat sterilization at 121°C and could be stored for at least 2 months at 40°C without significant change in particle size or particle size distribution.

Stability was lost, however, when F8H2PC was dispersed in a Hepes/NaCl buffer, or when a suspension of F8H2PC vesicles prepared in water was diluted using this buffer. It is known that the presence of electrolytes often destabilize vesicles made from single-chain surfactants. This paper reports that both shelf stability and the encapsulation stability of FnHmPC vesicles in the salt-containing Hepes-buffer can be restored by supplementing the fluorosurfactant with FnHm diblocks.

Materials and Methods.

The fluorinated surfactant F8H2PC and its hydrogenated analog H10PC were synthesized according to (*19, 20*). They were purified by column chromatography and repeated recrystallizations. Purity was assessed by TLC, ^{1}H, ^{19}F and ^{31}P NMR, HPLC and elemental analysis. The semi-fluorinated alkane F8H2 was a gift from Atochem (Pierre-Bénite, France). 5,6-carboxyfluorescein (CF), Hepes (*N*-2-hydroxyethyl-1-piperazine-*N'*-ethanesulfonic acid) buffer and Sephadex G-50 were purchased from Sigma. Decane came from Aldrich. Water for injectable preparations was used for all sample preparations.

Vesicles were prepared using a B-30 Branson sonicator (dial 4 on a 1-10 scale) fitted with a 3 mm-diameter titanium probe. Electron microscopy was performed with a Philips microscope (CM-12, 80 kV). Freeze fracture was effected with a Balzers cryofract (BAF 301, 2x10^{-6} torr). Vesicle mean diameters and size distributions were measured by laser light scattering (LLS) spectroscopy with a Coulter N4MD submicron particle analyzer (25°C). Cross polarization optical microscopy was performed with an Olympus BH2 microscope. Fluorescence measurements were made using a LS50B Perkin Elmer spectrofluorimeter equipped with a thermoregulated cell holder.

F8H2PC/F8H2 Vesicles. F8H2PC (25 mg), or an equimolar mixture of F8H2PC (12.5 mg) and F8H2 (9.0 mg), were solubilized in CHCl$_3$/CH$_3$OH (9:1) (0.5 mL) (total concentration of membrane components = 40 mM), and the solvent was removed by rotoevaporation. The resulting film was hydrated by the appropriate solution (water or Hepes buffer, containing CF or not) for 1 h at 40°C, then submitted to brief sonication (2 min) at the same temperature. A homogenous population of vesicles was formed as seen by LLS spectroscopy.

Carboxyfluorescein Entrapment and Release Experiments. Vesicles were prepared as described above in the presence of CF (100 mM, pH 7.5). They were annealed by incubation for 12h at 50°C in order to remove the defects present in the membrane as a result of sonication. Non-entrapped material was removed by chromatographying 60 μL of the dispersion of vesicles on a Sephadex G50 minicolumn, using a solution containing 20 mM Hepes buffer and 0.15 M NaCl as the eluent (*21*).

Release of the entrapped CF was monitored using fluorescence spectroscopy. This method exploits the self-quenching of the fluorescence of CF when confined at high concentration in the internal space of vesicles *(22)*. Release of CF through the bilayer membrane results in its dilution and consequent build-up of fluorescence. After removing the non-entrapped dye, 5 µL of the dispersion of vesicles were placed in a cuvette containing 2 mL of saline buffer. Fluorescence was measured at 25°C with an emission wavelength of 520 nm and an excitation wavelength of 480 nm. Maximum fluorescence was determined after lysis of the vesicles by 20 µL of sodium taurocholate (10% w/v). The time at which 80% of the probe is still encapsulated was directly measured on the curve.

Results

We have prepared vesicles both from F8H2PC alone and from a F8H2PC/F8H2 equimolar mixture, and we examined their stability in the Hepes/NaCl buffer by monitoring the evolution of their average diameter. We have also investigated the permeability of the vesicles by monitoring the release of entrapped 5,6-carboxyfluorescein (CF).

Vesicle Stability. We confirmed that vesicles made from F8H2PC alone, which are highly stable and heat sterilizable (121°C, 15 min) in water *(18)*, are not stable when prepared in Hepes/NaCl buffer. The dispersion precipitated within 1 hour suggesting the destruction of the vesicles. Freeze fracture electron micrographs of the precipitate showed the presence of large flat bilayer membranes; no vesicles were seen. Adding the Hepes/NaCl solution to dispersions of F8H2PC vesicles prepared in pure water also resulted in their destruction.

On the other hand, the vesicles made from the F8H2PC/F8H2 (1:1) mixture were found to be highly stable. Formation of a homogenous population of small unilamellar vesicles was observed. The vesicles' initial average mean diameter (140 ± 37 nm) remained essentially unchanged after 3 months at 25°C (152 ± 34 nm) (Figure 1). Polarization optical microscopy at 25°C of the coarse dispersion of the F8H2PC/F8H2 mixture obtained before sonication showed the typical defects of the fluid lamellar phase called « maltese crosses » (Figure 2). This suggests that the vesicles are in the fluid state at room temperature.

Membrane Permeability. We first examined the permeability of vesicles made from F8H2PC alone in Hepes/NaCl buffer and found that, in line with their instability in Hepes/NaCl buffer, these vesicles were unable to retain the entrapped carboxyfluorescein. The brown ring, typical of the migration of CF-containing vesicles on the Sephadex minicolumn, was not visible. This indicates that the release of the fluorescent probe was almost immediate, in any case too fast to be studied using our experimental procedure.

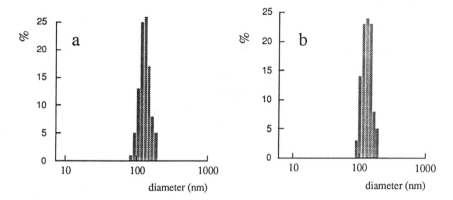

Figure 1. Diameter distribution of vesicles made from a mixture of a fluorinated single chain surfactant (F8H2PC) and a semi-fluorinated alkane (F8H2) in Hepes/NaCl buffer. Laser light scattering spectroscopy measurements were performed after preparation (a; 140 ± 37 nm) and after a 3 month storage at 25°C (b; 152 ± 34 nm).

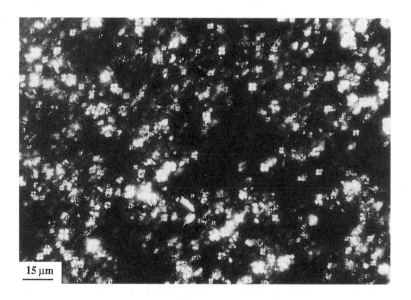

Figure 2. Cross-polarization optical microscopy of a coarse dispersion of non-sonicated F8H2PC/F8H2 vesicles. « Maltese crosses », which are typical defects of the fluid lamellar phase, can be seen.

By contrast, the F8H2PC/F8H2 vesicles prepared in the presence of CF and Hepes/NaCl buffer were found to be stable. The brown ring, characteristic of CF-loaded vesicles, was now observed on the minicolumn. The extrapolated half-life of the entrapped CF was around 2h ; a more accurate t_{80} value (time at which 80% of the entrapped probe is still present in the vesicles) was evaluated to 52 ± 6 min (Figure 3).

Aggregation Behavior of H10PC and H10PC/$C_{10}H_{22}$ (1:1) Mixtures. The hydrogenated analog of F8H2PC, H10PC, was seen to form micelles when dispersed in water. These micelles could not be visualized with freeze fracture electron microscopy. A dispersion of H10PC with $C_{10}H_{22}$ (decane) in 1:1 molar ratio was also prepared using the method utilized to produce the FnHmPC/FnHm vesicles. In that case an emulsion of decane in water was obtained, the phosphocholine derivative being the emulsifier. No evidence of formation of vesicles, hence no encapsulation of CF was observed (no brown ring on the minicolumn).

Discussion: a Molecular « Crutch » Effect

The Hepes/NaCl medium is commonly used when assessing applications of colloids as drug delivery systems. The fact that vesicles prepared from F8H2PC/F8H2 surfactant/diblock mixtures were stable in this solution, while those made from pure F8H2PC were destabilized by such a solution, can be rationalized in the following way.

It is well known that certain solutes and ions have a dehydrating effect on ionic and zwitterionic polar heads of phospholipids and can give rise to intermolecular forces that stabilize the more ordered phases (23). Their primary effect is to screen the net membrane surface charge. Although the effect of monovalent ions such as Na^+ is rather weak in the case of phospholipids, the situation could be quite different for single-chain surfactants for which crystallization of the hydrophobic chain can occur concomitantly with dehydration of the polar head. It is therefore reasonable to envision that, in the presence of the Hepes/NaCl medium, the F8H2PC vesicles could be in the gel state at room temperature. This could explain their lack of stability in Hepes/NaCl, since it is known that, in the absence of charges, gel-state vesicles are destabilized due to a decrease in the entropic repulsive forces (as compared to fluid vesicles) (24). This was observed with vesicles made from polyoxyethylene alkyl ether single-chain surfactants (25). On the other hand, when F8H2 is co-dispersed with F8H2PC, the diblock will have a strong tendency to remain in the bilayer membrane formed by F8H2PC. It was shown indeed that, due to their extreme hydrophobicity and to their lipophobicity as well, fluorinated chains tend to segregate and form an internal fluorinated film within the core of bilayer membranes (26).

As its molecular structure and size were chosen to correspond to those of the hydrophobic chain of F8H2PC, the diblock can be homogenously distributed in the bilayer. A schematic representation of the association of the two components is shown in Figure 4. The F8H2PC/F8H2 association can be considered, in the average, as being a reconstituted double-tailed entity, the diblock molecule acting like a « crutch » which strengthens the F8H2PC membrane. Being intercalated inbetween the F8H2PC chains, the diblock could thus hinder crystallization and therefore could maintain the F8H2PC/F8H2 vesicles in the fluid state at room temperature, which is supported by cross-polarization optical microscopy. The fact that decane did not induce the formation of a bilayer when co-dispersed with the micelle-forming hydrogenated analog H10PC, also evidences the structuring effect of the fluorinated chain.

Acknowledgements: The authors are grateful to Professor Jean G. Riess (MRI Institute, UCSD, San Diego CA) for helpful discussions. They also thank the Centre National de la Recherche Scientifique (CNRS) for support and ATOCHEM (Pierre Bénite, France) for the gift of fluorinated precursors.

References

1. Krafft, M. P.; Riess, J. G. *Biochimie* **1998**, *80*, 489.
2. Brace, N. O. *J. Org. Chem.* **1973**, *38*, 3167.
3. Turberg, M. P.; Brady, J. E. *J. Am. Chem. Soc.* **1988**, *110*, 7797.
4. Viney, C.; Russell, T. P.; Depero, L.E.; Twieg, R.J. *Mol. Cryst. Liq. Cryst.* **1989**, *168*, 63.
5. Höpken, J.; Möller, M. *Macromolecules* **1992**, *25*, 2482.
6. Russell, T. P.; Rabolt, J. F.; Twieg, R. J.; Siemens R. L.; Farmer B. L. *Macromolecules* **1986**, *19*, 1135.
7. Twieg, R. J.; Russell, T. P.; Siemens R.; Rabolt J. F. *Macromolecules* **1985**, *18*, 1361.
8. Riess, J. G.; Cornélus, C.; Follana, R.; Krafft, M. P.; Mahé, A. M.; Postel, M.; Zarif, L. *Adv. Exp. Med. Biol.* **1994**, *345*, 227.
9. Cornélus, C.; Krafft, M. P.; Riess, J. G. In Proceed. Int. Symp. Blood Substitutes. Vol 3. *The Fluorocarbon Approach*; Riess, J. G., Ed.; *Art. Cells, Blood Subst., Immob. Biotech.* **1994**, *22*, 1183.
10. Riess, J. G. In *Blood Substitutes*; Chang, T. M. S., Ed.; Landes Karger: New York, 1997, chap. 6, pp. 101-126.
11. Krafft, M. P.; Riess, J. G.; Weers, J. G. In *Submicronic Emulsions in Drug Targeting and Delivery*; Benita, S., Ed.; Harwood Academic Publ.: Amsterdam, 1998.
12. Krafft, M. P.; Dellamare, L.; Tarara, T.; Riess, J. G.; Trevino, L. Hydrocarbon oil/fluorochemical preparations and methods of use. PCT WO 97/21425.
13. Lattes, A.; Rico-Lattes, I. In Proceed. Int. Symp. Blood Substitutes. Vol 3. *The Fluorocarbon Approach*; Riess, J. G., Ed.; *Art. Cells, Blood Subst., Immob. Biotech.* **1994**, *22*, 1007.

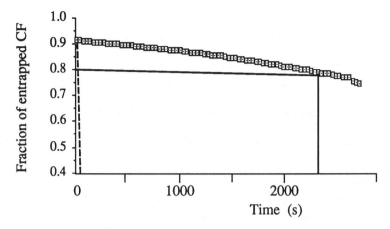

Figure 3. Carboxyfluorescein release from F8H2PC/F8H2 vesicles in NaCl/Hepes buffer. The release from vesicles made from F8H2PC alone is too fast to be measured and is represented by the dotted line (limit of the experimental method).

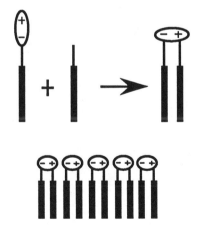

Figure 4. Schematical representation of the « crutch-like » association of F8C2PC and F8H2.

14. Trevino, L.; Frézard, F.; Rolland, J. P.; Postel, M.; Riess, J. G. *Colloids Surf.* **1994**, *88*, 223.
15. Trevino, L.; Frézard, F.; Postel, M.; Riess, J. G. *J. Liposome Res.* **1994**, *4*, 1017.
16. Krafft, M. P.; Ferro, Y. *Polymer Preprints* **1998**, 39, 938.
17. Privitera, N.; Naon, R.; Riess J. G. *Biochim. Biophys. Acta* **1994**, *1254*, 1.
18. Krafft, M. P.; Giulieri, F.; Riess, J. G. *Angew. Chem. Int. Ed. Engl.* **1993**, *32*, 741.
19. Krafft, M. P.; Vierling, P.; Riess, J. G. *Eur. J. Med. Chem.* **1991**, *26*, 545.
20 Sadtler, V. M.; Jeanneaux, F.; Krafft, M. P.; Rabai, J.; Riess, J. G. *New J. Chem.* **1998**, *22*, 609.
21 Frézard, F.; Santaella, C.; Vierling, P.; Riess, J. G. *Biochim. Biophys. Acta* **1994**, *1192*, 61.
22. Weinstein, J. N.; Yoshikami, S.; Henkart, P.; Blumenthal, R.; Hagins, W. A. *Science* **1977**, *195*, 489.
23. Seddon, J. M.; Cevc, G. In *Phospholipids Handbook*; Cevc, G., Ed.; Dekker, M.: New York, 1993; pp. 403-454.
24 Ostrowsky, N.; Sornette, D. In *Physics of Amphiphiles: Micelles, Vesicles and Microemulsions;* Degiorgio, V.; Corti, M., Eds.; Elsevier: Amsterdam, 1983.
25. van Hal, D. A.; Bouwstra, J. A.; van Rensen, A.; Jeremiasse, E.; De Vringer, T.; Junjinger, H. E. *J. Colloid Interface Sci.* **1996**, *178*, 263.
26. Riess, J. G. *J. Drug Targeting* **1994**, *2*, 455.

Chapter 5

Adsorption and Aggregation Properties of a Series of Semifluorinated, Unsaturated Fatty Acid Esters from Nonaqueous Solution

Richard R. Thomas[1], Jack R. Kirchner[1], and Douglas R. Anton[2]

[1]Jackson Laboratory, DuPont, Deepwater, NJ 08023
[2]Marshall Laboratory, DuPont, 3401 Grays Ferry Avenue, Philadelphia, PA 19146

A series of unsaturated, fatty acid esters of perfluorooctylethyl alcohol have been prepared. Surface tension isotherms and aggregation behavior were measured in n-butyl acetate solution. Analysis of the surface tension data using a Frumkin equation of state yielded thermodynamic parameters for adsorption. Degree of unsaturation or stereochemistry did not have much effect on the free energy of adsorption. These fluorinated esters were found to form small (~ 2 - 6) aggregates beginning at concentrations $> \sim 10^{-2}$ mol/L. A mass action analysis of the aggregation data was used to determine free energies of micellization. Degree of unsaturation and stereochemistry had a minor effect on aggregation thermodynamics and aggregate size.

The effect of unsaturation on the adsorption and aggregation properties of amphiphilic molecules has been studied in some detail. These studies have been performed in aqueous systems in which unsaturation is present on the hydrocarbon portion of the molecule. (1-3) This unsaturated hydrocarbon tail is the dominant feature at the air phase of the interface. These molecules also contain a hydrophilic head group (such as CO_2^-) that exists at the aqueous phase of the interface. The surface activity of fluorinated amphiphiles in aqueous solutions has been documented well also. (4-6) In this case, the fluorinated group is in excess at the air interface. The adsorption properties of semifluorinated alkanes in nonaqueous solutions have, in addition, been studied. However, there have been no studies of the effect of unsaturation on the hydrocarbon portion of semifluorinated amphiphiles in nonaqueous systems. The current study will examine the adsorption and micellization behavior of a series of semifluorinated, unsaturated fatty acid esters in nonaqueous solution.

It is known well from the study of fatty acid esters by the film balance technique, that behavior (surface pressure and molecular area) is influenced not only by the type of functional group at the surface, but its orientation as well. (7) Surface pressure/area isotherms can be used to measure the surface area occupied by a molecule at its collapse pressure. For example, molecular areas at collapse for stearic and oleic acids cast on water yield ~ 20 and ~ 30 Å^2, respectively. (7) The larger area of the oleic acid is a consequence of the "kink" in its molecular structure caused by the presence of the *cis* double bond.

Fluorinated fatty acids adsorbed to the air interface have been studied also. (8) It was found that there is a direct relationship between number of -CF_2- groups in the chain to the film pressure at collapse and molecular area. Under similar conditions in the subphase, the fluorinated materials tend to have molecular areas at collapse that are ≈ 10 - 40 % larger than the hydrocarbon analogs. This can be explained easily by cross-sectional area differences of the respective chains and the effect on packing efficiency at the surface. (9)

However, few studies have been performed in nonaqueous solutions. Furthermore, most of the work done previously has used molecules with rather simple hydrophilic groups (e.g., -CO_2H or -CO_2^-). No studies have examined the effect of orientation of the hydrocarbon chain in the *liquid subphase*. In simple aqueous systems, the hydrocarbon portion of the molecule is always the hydrophobe and, therefore, in the air phase. The systems studied here are just the opposite. The hydrocarbon portion of the molecule is, relatively, more hydrophilic than the fluorocarbon portion. The study of this effect on various system properties is the subject of this investigation.

Experimental

Materials. Elaidic (98%), linoleic (99%) and oleic (99+%) were from Aldrich Chemical Co. and used as received. *N*-butyl acetate (99+%, ACROS Organic Chemicals, Spectral Grade) as used as received. Perfluorooctylethyl alcohol ($F(CF_2)_8(CH_2)_2OH$) was from Clariant (> 99% C_8F_{17} telomer by GC).

Preparation of the Linoleic Acid Ester of Perfluorooctylethyl alcohol. A three-necked 100 mL round-bottomed flask fitted with nitrogen inlet/outlet, thermometer, paddle stirrer and a short-path distillation head, condenser and receiver is charged with 24.94 g (88 mmol) of linoleic acid, 45.76 g (99 mmol) of perfluorooctylethyl alcohol and 0.12 g of 70 % aqueous H_3PO_3. The reaction mass was heated with stirring to 140-150° C for ~ 48 hours. The sample was allowed to cool under positive nitrogen pressure. The reaction mass yielded 58.02 g (91 % yield based on linoleic acid) of perfluorooctylethyl linoleate as a pale yellow liquid: mp -9.2°C; ^{1}H NMR δ 0.2-2.1 (-CH_2-, 23H); 2.2-2.4 (-CH_2-, t, 2H); 2.5-2.8 (-CH_2-, t of t, 2H); 2.7-2.9 (-CH_2-, t, 2H); 4.3-4.5 (-CH_2-, t, 2H); 5.2-5.5 (=CH-, m, J ≈ 10 Hz, 4H).

Anal. Calcd for $C_{28}H_{35}O_2F_{17}$: F, 44.5. Found: 43.3. Product esters were confirmed by GC-MS analysis of the reaction mass. The other esters were prepared in similar fashion.

Surface Tension Measurements. Surface tension measurements were performed by the pendant drop method using a VCA-2500 contact angle goniometer from AST Products (Billerica, MA). The device has a CCD camera, frame grabber and software which does the pendant drop calculation by curve fitting. The drop was allowed to equilibrate for ~ 15 seconds before measurement. It is necessary to input solution density values to solved for surface tension. These were measured using a Mettler/Paar DMA46 densiometer. Solutions were prepared by dilution of the pure acid esters of the respective fluorinated alcohol in *n*-butyl acetate. The measurements were conducted at 25°C. Solutions were prepared just prior to use to prevent air oxidation.

Vapor Pressure Osmometry. Measurements were performed using a Wescan Model 233 vapor pressure osmometer (Jupiter Instrument Co.). The instrument was calibrated using squalene (Eastman; purified by fractional crystallization) in *n*-butyl acetate. The operating temperature during data collection was 81.0 ± 0.1 °C.

Results and Discussion

Synthesis of Semifluorinated, Unsaturated Fatty Acid Esters. The amphiphiles used in the current study are based on the phosphorus acid catalyzed condensation of perfluorooctylethyl alcohol and a C_{18} unsaturated fatty acid. The reaction scheme for preparation of the elaidic acid ester of perfluorooctylethyl alcohol (EAE) is shown in Figure 1. Similar methods were used to prepare the other esters. The oleic and linoleic acid ester derivatives are designated OAE and LAE, respectively. The semifluorinated esters were prepared in high yields and good purity. The purity of the compounds was verified by GC and ^1H and ^{19}F NMR. Space-filling models of the esters prepared (elaidic, oleic and linoleic) are shown in Figure 2.

Surface Tension Isotherms. A study was made of the surface tension isotherms of several fluorinated esters prepared using $F(CF_2)_8(CH_2)_2OH$. Shown in Figure 3 are the data for EAE, LAE and OAE in *n*-butyl acetate solution. While the shapes of the curves are similar, there are some subtle differences which may be discerned through proper theoretical analysis. At the high concentration limit, surface tension values did approach those of the pure compounds (OAE = 18.3 ± 0.1 and LAE = 19.2 ± 0.1 mN/m; EAE is a solid at room temperature).

The method chosen to analyze the data was based on the Frumkin equation of state (2,10)

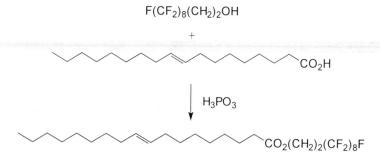

Figure 1. Synthesis of elaidic acid ester of perfluorooctylethyl alcohol.

Elaidic acid ester of perfluorooctylethyl alcohol

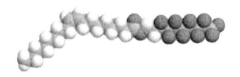

Oleic acid ester of perfluorooctylethyl alcohol

Linoleic acid ester of perfluorooctylethyl alcohol

Figure 2. Space-filling models for semifluorinated, unsaturated fatty acid esters. $-(CF_2)_8F$ portion of molecule to right.

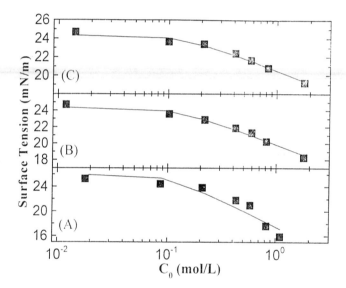

Figure 3. Surface tension isotherm data for elaidic acid (A), oleic (B) and linoleic acid (C) esters of $F(CF_2)_8(CH_2)_2OH$. Solid lines are data fit according to Frumkin adsorption isotherm (eq 1).

$$\pi = -RT\Gamma_{\infty}[\ln(1-\theta)+(H^S/RT)\theta^2]$$ (1)

where $\pi = \gamma^0 - \gamma^S$, γ are the surface tensions of pure solvent (0) and solution (S) (N/m), R is the Boltzmann constant (J mol^{-1} K^{-1}), T is temperature (K), Γ and Γ_{∞} refer to surface excess at concentration C_0 and at saturation (mol/m^2), respectively, $\theta = \Gamma/\Gamma_{\infty}$ (surface coverage at π), and H^S is the partial molar free energy of surface mixing of surfactant and solvent (J/mol). The quotient θ is not accessible readily from experimental data, so an equation relating another experimental quantity, the concentration of surface active species, C_0 (mol/L) to θ is needed.

$$C_0 = \frac{a\theta}{1-\theta}\exp\left[\frac{-2H^S}{RT}\theta\right]$$ (2)

with

$$a = \exp(\Delta G^0/RT)$$ (3)

where ΔG^0 is the standard free energy of adsorption (J/mol) and is given by (11)

$$\Delta G^0 = -RT\ln(\pi/\chi)_{\chi\to 0}$$ (4)

where χ is the mole fraction of surface active species. The quantity ΔG^0 is found graphically by a plot of ΔG^0 vs χ and extrapolating to infinite dilution.

Equations 1 and 2 must be solved simultaneously since they both contain a common term, H^S. The strategy employed involves simultaneous nonlinear least squares curve fitting of equation 1 with a root finder to find the value of θ that corresponds to the measured quantity C_0. The method of choice for doing nonlinear least squares curve fitting is based on the algorithm of Marquardt and Levenberg. (12) An algorithm based on Brent's method to solve for the roots of equation 2 was incorporated after each iteration of curve fitting. (13)

Shown in Figure 3 are surface tension isotherm data for elaidic acid (A), oleic (B) and linoleic acid (C) esters of $F(CF_2)_8(CH_2)_2OH$ along with curves (solid lines) fit according to the Frumkin adsorption isotherm (equations 1 and 2). The thermodynamic parameters obtained from solution of the Frumkin adsorption isotherm are given in Table I.

As expected, the standard free energy of adsorption, ΔG^0, for all systems are comparable and < 0. This implies that the nature of the double bonds in the fluorinated esters do not seem to play an important role in surface activity. In contrast, much larger stereochemical effects are seen with surface active hydrocarbons in aqueous solution. (2) ΔG values for adsorption tend to be more negative for *trans* compared to *cis* isomers. (1-3) The free energy gain associated with adsorption can be attributed to rearrangement of solvent molecules during the transfer of solute from solution to the air interface. In this system, it appears that *n*-butyl acetate is a relatively good solvent and there may not be a very great energetic gain based simply on rearrangement of solute around hydrocarbon chains of various configurations.

Table I. Thermodynamic Parameters for Semifluorinated, Unsaturated Fatty Acid Solutions.

Ester [a]	ΔG^0 (kJ/mol)	H^S (kJ/mol)	$\Gamma_\infty \times 10^{-7}$ (mol/cm^2)	Molecular area ($\mathring{A}^2$/molecule)	χ^2 [b] $\times$ 10^{-8}	s (mN/m) [c]
Elaidic acid	-9.44 ± 0.18	7.93 ± 0.04	15.7 ± 0.02	105.6 ± 1.5	49	0.70
Linoleic acid	-9.18 ± 0.21	5.42 ± 0.13	8.89 ± 0.28	186.7 ± 5.9	2.6	0.16
Oleic acid	-9.93 ± 0.42	5.68 ± 0.15	9.52 ± 0.26	174.3 ± 4.7	3.0	0.17

[a] Prepared using $F(CF_2)_8(CH_2)_2OH$.

[b] $\chi^2 = \sum_{i=1}^{N} (\pi_{obs} - \pi_{calc})^2 / (N-n)$, where N is the number of data points and n is the number of parameters being fitted (N-$n \equiv$ degrees of freedom).

[c] Mean error in surface pressure, π.

Solutions of the fluorinated, unsaturated fatty acid esters clearly form nonideal solutions as judged by values of $H^S \neq 0$. The magnitude of H^S can be used to estimate the interaction of the various fluorinated esters at the interface. In all cases H^S is relatively small and should not be surprising in light of the limited solubility of fluorocarbons in hydrocarbon solvents. The value of H^S will have a noticeable impact, *vide infra*, on molecular areas. (14). For comparison, an attempt to examine the π - C_0 data using the Frumkin isotherm with $H^S = 0$ (Langmuir adsorption isotherm), led to much worse fits statistically and unreasonably large molecular areas.

More striking differences are seen in the molecular area at surface saturation determinations. Within statistical error, the molecular area values for LAE and OAE are comparable at 187 ± 6 and 174 ± 5 Å^2 compared to the value of EAE at 106 ± 2 Å^2. As mentioned previously, film balance techniques have been used to determine molecular area demands at film collapse pressures for hydrocarbon, fatty acids in aqueous subphases. Areas determined are in the range 20 -30 Å^2/molecule. (7) Using similar techniques, molecular areas at film collapse have been determined for fluorinated, fatty acids on water. (8) These values are in the range 20 - 40 Å^2/molecule. Clearly, the fluorinated, unsaturated fatty acid esters employed in the current study have surface molecular areas which are in excess of theoretical predictions based on monomeric species. Large limiting surface areas (> cross-sectional area of perfluoroalkyl chain $\approx$ 30 Å^2) have also been measured during adsorption of semifluorinated alkanes at nonaqueous/air interfaces. This was explained as due to the presence of a relatively expanded film of fluorocarbon at the interface. This explanation can be used in the current study, *vide infra*, reasonably. (15)

Osmometric Studies of Aggregation. Surface molecular areas of the fluorinated, unsaturated fatty acid esters used in the current study prompted a search for aggregation phenomena. Solutions of EAE, LAE and OAE were prepared in *n*-butyl acetate solvent and examined by vapor pressure osmometry at 81°C. The results are shown in Figure 4. All of the fluorinated, unsaturated fatty acid esters exhibit substantial degrees of aggregation under the current experimental conditions, as judged by a slope < 1 from the plots of $C_{osmotic}$ vs $C_{stoichiometric}$.

These types of molecules are prone to air oxidation. However, ideal osmometric behavior at low concentrations verified that oxidation, especially leading to oligomerization, had not occurred.

Using osmometry data, number average aggregation values can be calculated (as $[C]_{stoichiometric}/[C]_{osmotic}$). Shown in Figure 5 are $[C]_{osmotic}$ and number average aggregation values for EAE (A), OAE (B) and LAE (C) in *n*-butyl acetate solution, respectively.

Using aggregation values, standard free energies of micellization, ΔG^o_{mic}, can be calculated. (16)

$$\Delta G^0_{mic} = - RT \ln \beta \qquad (5)$$

The parameter β arises from a mass action analysis of aggregation formation constants. Therefore, β is a product of all the stepwise aggregation constants up to n.

$$\beta = [C_n][C_1]^{-n} \qquad (6)$$

Where $[C_n]$ is the concentration of monomers consumed to form an aggregate of size n and $[C_1]$ is the concentration of monomers which remain. Shown in Table II are values of ΔG^o_{mic},

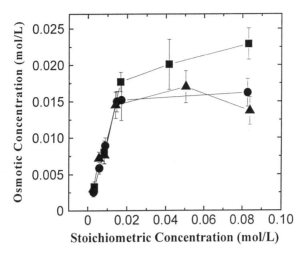

Figure 4. Stoichiometric vs osmotic concentrations for elaidic (●), linoleic (■) and oleic acid esters (▲) of $F(CF_2)_8(CH_2)_2OH$.

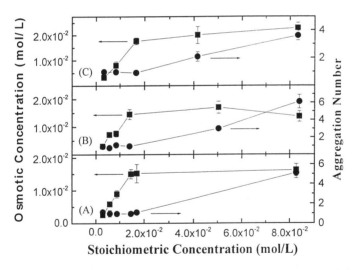

Figure 5. Osmotic concentration and aggregation numbers of elaidic (A), oleic (B) and linoleic (C) acid esters of $F(CF_2)_8(CH_2)_2OH$.

Table II. Thermodynamic Data for Aggregation of Semifluorinated, Unsaturated Fatty Acid Esters.

Sample	ΔG^0_{mic} (kJ/mol)	Number Average Aggregation Value	[M] (mol/L) [a]
EAE	-24.0 ±3	5.10 ± 0.5	8.62×10^{-2}
LAE	-15.9 ± 2	3.63 ± 0.3	8.31×10^{-2}
OAE	-30.0 ± 5	6.12 ± 0.8	8.38×10^{-2}

[a] Concentration at which deviation of osmometric data from ideal monomer behavior is first exhibited.

aggregation values and concentrations at which they were obtained.

It is important to define the concentration at which calculations are made since surface active molecules in nonaqueous solution often do not possess a true critical micelle concentration (CMC) as do aqueous surfactant systems. Hence, the use of equations 5 and 6. This is due to the stepwise formation of aggregates as a function of concentration as opposed to the production of rather discrete, monodisperse micelles at the CMC in aqueous solution. (17). The similarities in deviation from ideal stoichiometric concentration behavior (~ 8 × 10^{-2} mol/L) for all of the amphiphiles examined indicate that degree of unsaturation in the hydrocarbon tail and stereochemistry have little effect on the concentration at which aggregation begins. Contrast this to the behavior of simple unsaturated hydrocarbon surfactants in aqueous solution. The presence of the a double bond in the hydrocarbon tail is observed to increase the CMC by a factor of 2 - 3. (18-20) If micellization is dominated by entropic forces due to solvent reorganization, (21) micellization in the current system would be expected to be nearly invariant with degree of unsaturation and stereochemistry. This can be rationalized by less solvation and, hence, less reorganization in the current system than in an aqueous one.

The trend in - ΔG^o_{mic} (OAE ≥ EAE > LAE) follows the same as the decrease in melting points of the derivatives (EAE, OAE and LAE = 30.5, 12.7 and -4.5°C, respectively) indicating a higher propensity for interaction based on packing efficiency of the molecule. Not many studies have been performed which study aggregates of fluorinated materials in nonaqueous systems, so a comparison of ΔG^o_{mic} values obtained here with other investigations is impossible. However, ΔG^o_{mic} values measured for aqueous systems and hydrocarbon surfactants are often more negative. This should not be surprising based on the relative solubilities of materials.

In a somewhat related series of studies by Kunitake et al, the solution and adsorption properties of a variety of double-chain perfluoroalkyl amphiphiles were investigated. (22-25) The derivatives included an oleic amide in the hydrocarbon tail. The authors claim that the main driving force for aggregation was not entropic in origin but, rather, due to cohesive energy differences between solute and solvent. However, these double-chain perfluoroalkyl amphiphiles formed bilayers that exhibit a bilayer to monomer conversion at elevated temperatures (70°C) by VPO and differential scanning calorimetry measurements. (25) From thermodynamic measurements, ΔH and ΔS of the monomer to bilayer aggregation process were evaluated as -70 kJ/mol and -230 J/(mol $\bullet$ K), respectively. In the present system, there was no evidence seen for bilayer formation and the aggregates were stable with a relatively constant aggregation number over a wide temperature range. This would account for the values of ΔH and ΔS observed in the current study. It is also interesting to note that the adsorption isotherm for the oleic amide derivative used in the previous study (25) exhibited a concave upward curve indicating stepwise adsorption and prevented a molecular area calculation. In the present study, all the fluorinated derivatives show downwardly concave surface tension isotherm curves that are typical of a simple phase diagram for adsorption predicted by the Gibbs adsorption isotherm.

Comparing the molecular areas, Table I, determined by measuring surface tension isotherms and using the Frumkin equation of state (equations 1 and 2) with aggregation values in Table II, it is clear that aggregation is occurring and this has an influence on solution surface tension properties. A previous study has provided evidence that aggregation might persist into the solid state. Atomic, friction and force modulated microscopic techniques have been used to examine surfaces of coatings prepared with these semifluorinated, unsaturated fatty acid esters. (26) The images are, indeed, heterogeneous chemically with domains of fluorochemicals existing in a sea of hydrocarbon coating in a situation resembling a binodal phase-separated state. Interestingly enough, the force microscopic images indicate that the fluorinated domains are on the order of ~ 300 Å in diameter. This diameter is similar to molecular areas (Table I) estimated from solution of the Frumkin isotherm equation using surface tension data.

In addition, in a separate study from this laboratory, variable temperature ^{19}F NMR was used to verify the aggregation observed by vapor pressure osmometry. The results obtained were similar to those from osmometric studies. (27) Furthermore, analysis of the surface tension isotherm data shown in Figure 3 using the Gibbs adsorption isotherm equation yielded molecular areas at saturation that were comparable to those obtained using the Frumkin adsorption isotherm. (27)

The quotient of molecular area (Table I) and respective aggregation value (Table II) allows for a more accurate estimate of true molecular area. Such a calculation yields 21, 51 and 29 Å^2/molecule for elaidic, linoleic and oleic acid esters, respectively. This is a simplistic estimate based on disk-like aggregates at the surface. These areas are remarkably close to those measured for fluorinated acids in aqueous solution using Langmuir-Blodgett techniques. (8) Moreover, the trend in molecular area, as a function of degree of unsaturation and stereochemistry, follows that observed with surface active hydrocarbon molecules. This implies that the molecules exist as aggregates at the liquid surface with the fluorocarbon chain oriented towards the air interface. The difference in molecular area can, perhaps, be attributed to the volume occupied by the *hydrocarbon* portion of the molecule in the *liquid* phase. This can be seen by examination of the space-filling models shown in Figure 2. Place the perfluoroalkyl group orthogonal to the surface and note the volume displaced by the hydrocarbon tail upon rotation along $F(CF_2)_8$- axis. It can be imagined easily that the hydrocarbon tail will be become the dominant factor in packing at the interface when unsaturation is present.

While few studies have been conducted for fluorinated, surface active agents in nonaqueous solution, those that have also find aggregation occurring at relatively low concentrations. (28) In addition, the aggregate numbers are low when compared to similar systems in aqueous solution. (28) Evidence suggests that the aggregates formed have the fluorocarbon portion of the molecule oriented towards the core. (29)

Conclusions

Semifluorinated ester derivatives of a variety of unsaturated fatty acids can be prepared easily and in high yield by reaction with $F(CF_2)_8(CH_2)_2OH$. All of the materials prepared have been shown to be quite surface active in nonaqueous solution. The presence of unsaturation in the hydrocarbon chain of the amphiphilic molecules and stereochemistry seem to have little effect on the degree of aggregation and ΔG for adsorption at the solvent/air interface. Molecular areas, determined from fits of the Frumkin equation of state, to surface tension isotherm data, are larger than would be expected based on cross-sectional areas of perfluoroalkyl chains organized perpendicularly towards the air interface. This is attributed to small aggregates that persist into the solid state. When measured molecular areas are divided by degree of aggregation, reasonable molecular areas (21-51 Å^2/molecule) are obtained.

Literature Cited

(1)   Burczyk, B.; Piasecki, A.; Weclas, L., *J. Phys. Chem.* **1985**, *89*, 1032.

(2)   Lunkenheimer, K.; Burczyk, B.; Piasecki, A.; Hirte, R., *Langmuir* **1991**, *7*, 1765.

(3)   Piasecki, A.; Sokolowski, A.; Burczyk, B.; Gancarz, R.; Kotlewska, U., *J. Colloid Interface Sci.* **1997**, *192*, 74.

(4)   Lin, I. J., *J. Phys. Chem.* **1972**, *76*, 2019.

(5)   Janczuk, B.; Méndez Sierra, J. A.; González-Martín, M. L.; Bruque, J. M.; Wójcik, W., *J. Colloid Interface Sci.* **1997**, *192*, 408.

(6)   Achilefu, S.; Selve, C.; Stébé, M.-J.; Ravey, J.-C.; Delpuech, J.-J., *Langmuir* **1994**, *10*, 2131.

70

(7)   Hann, R. A. in *Langmuir-Blodgett Films*; Roberts, G., Ed.; Plenum: New York, 1990; Chapter 2.

(8)   Bernett, M. K.; Zisman, W. A., *J. Phys. Chem.* **1963**, *67*, 1534.

(9)   Hoernschemeyer, D., *J. Phys. Chem.* **1966**, *70*, 2628.

(10)   Lin, S.-Y.; McKeigue, K.; Maldarelli, C., *Langmuir* **1991**, *7*, 1055.

(11)   Sokolowski, A.; Piasecki, A.; Burczyk, B., *Langmuir* **1992**, *8*, 1775.

(12)   Press, W. H.; Flannery, B. P.; Teukolsky, S. A.; Vetterling, W. T. *Numerical Recipes (The Art of Scientific Computing)*; Cambridge University Press: New York, 1986; Chapter 14.

(13)   Press, W. H.; Flannery, B. P.; Teukolsky, S. A.; Vetterling, W. T. *Numerical Recipes (The Art of Scientific Computing)*; Cambridge University Press: New York, 1986; Chapter 9.

(14)   Lucassen-Reynders, E. H., *J. Colloid Interface Sci.* **1973**, *42*, 554.

(15)   Binks, B. P.; Fletcher, P. D. I.; Sager, W. F. C.; Thompson, R. L., *Langmuir* **1995**, *11*, 977.

(16)   Kertes, A. S. in *Micellization, Solubilization, and Microemulsions*; Mittal, K. L., Ed.; Plenum Press: New York, 1977; Vol. 1; pp 445-454.

(17)   Phillips, J. N., *Trans. Faraday Soc.* **1955**, *51*, 561.

(18)   Sprague, E. D.; Duecker, D. C.; Larrabee, C. E., Jr., *J. Colloid Interface Sci.* **1983**, *92*, 416.

(19)   Durairaj, B.; Blum, F. D., *J. Colloid Interface Sci.* **1985**, *106*, 561.

(20)   Larrabee, C. E., Jr.; Sprague, E. D., *J. Colloid Interface Sci.* **1986**, *114*, 256.

(21)   Nusselder, J. J. H.; Engberts, J. B. F. N., *J. Colloid Interface Sci.* **1992**, *148*, 353.

(22)   Kunitake, T.; Okahata, Y.; Shimomura, M.; Tasunami, S.-i.; Takarabe, K., *J. Am. Chem. Soc.* **1981**, *103*, 5401.

(23)   Ishikawa, Y.; Kuwahara, H.; Kunitake, T., *Chem. Lett.* **1989**, 1737.

(24)   Kuwahara, H.; Ishikawa, Y.; Kunitake, T., *Chem. Lett.* **1993**, 1161.

(25)   Ishikawa, Y.; Kuwahara, H.; Kunitake, T., *J. Am. Chem. Soc.* **1994**, *116*, 5579.

(26)   Sauer, B. B.; McLean, R. S.; Thomas, R. R., *Langmuir* **1998**, *14*, 3045.

(27)   Thomas, R. R.; Glaspey, D. F.; DuBois, D. C.; Kirchner, J. R.; Anton, D. R.; Lloyd, K. G.; Stika, K. M., *Langmuir* **2000**, *16*, 6898 .

(28)   Fowkes, F. M. in *Solvent Properties of Surfactant Solutions*; Shinoda, K., Ed.; Marcel Dekker: New York, 1967; Vol. 2; Chapter 3.

(29)   Duns, G. J.; Reeves, L. W.; Yang, D. W.; Williams, D. S., *J. Colloid Interface Sci.* **1991**, *145*, 270.

Chapter 6

Wetting Behavior of Thin Films of Polymethacrylates with Oligo(hexafluoropropene oxide) Side Chains

Sergei S. Sheiko, Peter-Jan Slangen, Maarten Krupers[1], Ahmed Mourran, and Martin Möller

Organische Chemie III/Makromolekulare Chemie, Universität Ulm, 86069 Ulm, Germany

Polymethacrylates modified with non-crystallizable oligo(hexafluoro-propene oxide) side chains were investigated regarding their wetting of a mica substrate. Polymers displayed pseudo partial wetting, when the original film ruptured and the opened area remained covered by a thin layer of the material. Depending on the number of HFPO units in the side chain, the dewetting resulted in different film structures. The pentamer substituted PMA underwent microphase separation and, as a consequence, displayed so-called autophobic wetting when micrometer sized droplets emerged on an oriented monolayer of the same material. In the case of the trimer substituted PMA, the homogeneous films were stable above a certain film thickness determined by the chain elasticity.

Low surface energy and low adhesion to both aqueous and organic media are the most characteristic and valuable properties of fluorocarbon coatings (1). Maximum oil and water repellency is observed upon increasing the concentration of trifluoromethyl (-CF_3) end groups at the free surface (2-4). So far most studies have focused on linear polymers containing perfluoroalkyl side chains (4-10). In this case, microphase separation results in a bilayer morphology composed of alternating layers of the fluorocarbon pendants and the hydrocarbon backbone. Due to the outward orientation of the -CF_3 groups, critical surface tensions as low as 6 mN/m were obtained for the homopolymer of p-perfluorooctyl-ethylenoxymethyl styrene (6). In other words, there is no organic, metalo-organic and even fluoro-organic liquid which could wet such a surface. Moreover, the surface energy could be smoothly varied from 9 to 40 mJ/m^2 by copolymerization with nonfluorinated monomers (11).

Apart from the surface activity of the -CF_3 groups, other factors can affect the structure and surface properties of semifluorinated polymers. Microphase separation of

[1]Current address: TNO Industrie, P.O. Box 6031, 2600 JA Delft, The Netherlands

the hydrocarbon backbone and the fluorinated side chains affect the conformation of the polymer chain and results in a significant entropic penalty (12-14). Hereby, both the mixing and configurational entropy as well as characteristic relaxation times depend on the degree of polymerization.

For thin films, also short range specific interactions and a variation of the long range surface forces with the film thickness have to be taken into account (15,16). Recently, we have reported on the competition between the interfacial interactions and disordering of the polymer backbone, causing self-dewetting of perfluoroalkyl substituted PMMA on glass (17). In spite of the very low surface energy, the spincast films turned out to be unstable upon heating above the isotropization temperature $T_i =$ 90°C. The proposed explanation of the self-dewetting process was based on the ex-situ SFM measurements of the dewetting pattern which was quenched by cooling the sample down to room temperature. The multilayer structure ruptured and coagulated into droplets via a peculiar stepwise, i.e., layer-by-layer dewetting process. The origin of this phenomenon was explained by disordering of the top layer, which surface energy became somewhat higher compared to the ordered sublayer. The surface tension of the isotropic melt of F8H2-PMA was measured to be 17 mN/m, which is significantly higher than the critical surface tension $\gamma_c = 8$ mN/m determined for a spincast film. This difference in the surface energy between a disordered melt phase and an oriented sublayer was considered to be a cause of the film instability and formation of the droplets. The stepwise manner of the film melting might be connected with a different stability of the layers depending on their proximity to a hard wall of the substrate surface.

Thus, the wetting behavior of such polymers was affected by smectic ordering of the side chains. In this case, phenomena like surface freezing (18,19), vitrification of a heterogeneous structure, and also solvent entrapping, become increasingly important for formation and stability of polymer films containing crystallizable groups. Recently, interfacial ordering of low molecular weight surfactants has been shown to cause an unusual layer-thinning behavior near the isotropization transition (20).

In this paper we describe an approach to exclude side chain crystallization effects and to approach equilibrium conditions for wetting of a solid surface by fluid amphiphilic polymers. A series of polymethacrylates (PMA) containing oligo(hexafluoropropene oxide) side chains was synthesized. The hexafluoropropene oxide oligomers cannot crystallize because of (i) the segmental flexibility enhanced by the in-chain oxygen atoms and (ii) the irregular configuration of the chiral C-atoms. Moreover, multiple CF_3 groups per monomer can be expected to improve the surface activity compared to linear perfluoroalkyl side chains. Wetting properties of the prepared polymers were investigated by scanning force microscopy, which enables visualization of the submicrometer structure of the liquid films.

Polymers and Material Properties

Synthesis. The synthesis of the three different oligo(HFPO)-containing methacrylates, and their radical polymerization in solution have been described elsewhere (21,22). The length of the oligo(HFPO) units was varied from trimer, via tetramer to pentamer. The chemical structure of the oligo(HFPO) substituted polymethacrylates is shown in the scheme below.

The polymers were characterized by size exclusion chromatography (SEC) in 1,1,2-trichlorotrifluoroethane (freon-113) using well defined polyisoprenes as standards. Thus determined relative molecular weights are depicted in Table 1.

Phase Behavior. At room temperature, the prepared polymers were clear, viscous materials. The phase behavior of the polymers was investigated by differential scanning calorimetry (DSC) and optical microscopy between crossed polarizers. DSC was performed with a DSC-7 differential scanning calorimeter from Perkin Elmer. The low temperature measurements were calibrated with cyclopentane, cyclohexane, water, gallium and indium. For all samples neither DSC-traces nor thermooptical investigations rendered any evidence for crystallization effects. The polymer with the HFPO trimer side chains shows one glass transition (Table 1), whereas the presence of tetramer and pentamer side chains results in two glass transitions.

Table 1. Molecular weight, degree of polymerization, and glass transition temperatures of P(HFPO$_n$MA) homopolymers.

Polymer	M_w, (g/mol)[a]	M_w/M_n	$T_{g,1}$ (°C)[b]	$T_{g,2}$ (°C)[b]
P(HFPO$_3$MA)	$1.8 \cdot 10^5$	5.2	-8	---
P(HFPO$_4$MA)	$5.1 \cdot 10^4$	1.7	-34	58
P(HFPO$_5$MA)	$2.5 \cdot 10^4$	3.2	-47	73

[a] Molecular weight was determined by SEC in 1,1,2-trichlorotrifluoroethane relatively to polyisoprene standards.
[b] Glass transition temperatures were determined by the point of inflection in the heat capacity step.

The samples which gave two transitions were optically clear and homogeneous as it is typical for microphase separation. Presumably, a HFPO rich and a PMA rich phase were formed giving rise to a lower and higher glass transition temperature, respectively.

Structure and Surface Properties of Thin Films

Surface Properties. Since the oligo(HFPO) polymers contain more than one -CF$_3$ group per monomer unit, each side chain can contribute several CF$_3$ groups for adsorption at the free surface. The improved structural composition might balance eventual worsening of the surface properties caused by the disordered molecular structure. Contact angles were measured by means of the sessile drop method (G40, Krüss GmbH). at a constant temperature of 20±0.1 °C. A homologous series of *n*-alkanes (octane, decane, dodecane and hexadecane) was used as wetting liquid. Corresponding contact angles are depicted in Table 2

Table 2. Contact angles P(HFPO$_n$MA) homopolymers measured by the sessile drop technique for different test liquids

Polymer	*C16*[a]	*C12*[b]	*C10*[c]
P(HFPO$_3$MA)	74.8	66.3	57.2
P(HFPO$_4$MA)	74.6	74.3	71.8
P(HFPO$_5$MA)	70.6	n.r.[d]	n.r.[d]

[a] hexadecane, γ= 27.7 mN/m, [b] dodecane, γ= 25.4 mN/m, [c] decane, γ= 24.1 mN/m
[d] the data were not reproducible

The Good-Girifalco-Fowkes-Young (GGFY) equation was applied for the calculation of the dispersion force surface energy, γ_s^D:

$$\cos\theta = -1 + 2(\gamma_s^D)^{1/2}(\gamma_L)^{-1/2}$$

where γ_L is the surface tension of the wetting liquid. This approach is valid for n-alkanes, as the polar contribution to the work of adhesion is negligible. When $\cos\theta$ is plotted as a function of $(\gamma_L)^{-1/2}$ the dispersion force surface energy can be obtained by extrapolation to $\cos\theta = 1$.
All three polymers showed dispersion force surface energy γ_s^D= 11 ± 1 mN/m, which is markedly lower than 17 mN/m found for perfluoroalkyl substituted PMA in the melt. In contrast to perfluoroalkyl substituted PMA, that showed higher contact angles after annealing, the surface properties of the oligo(HFPO) polymer films were invariant. This effect was expected due to the high flexibility of the fluoroether segments allowing already a high degree of surface segregation without an additional temperature treatment.

Thin Film Morphology. The film morphology was investigated by scanning force microscopy (SFM) which was already shown before to enable structure visualization of viscous liquids (*23*). SFM was performed at ambient conditions with a Nanoscope III (Digital Instruments) operated in the tapping mode at a resonance frequency of ~ 360 kHz. Si probes with a spring constant of ~ 50 N/m were employed. Samples for the SFM measurements were prepared by spincasting a solution of the polymer in 1,1,2-trichlorotrifluoroethane (freon-113) on a freshly cleaved mica crystal. The concentration was varied from 0.1 to 1 g/L in order to control the coverage of the substrate. Measurements were done directly after spincasting and then after long term annealing at room temperature.

In the following the film structure will be compared between the HFPO trimer and pentamer substituted polymers, as they display a significant difference regarding phase separation (Table 1). The tetramer substituted PMA showed a wetting behavior similar to the pentamer sample. Figure 1 shows two SFM micrographs of the films, prepared by spincasting of a c = 1 g/L solution. Both samples demonstrate a complete coverage of the substrate by a thin film, the thickness of which was measured to be about 12 nm. The small holes (black areas) in the center of the image in Figure 1a emerged after scanning the 3x3 μm^2 area before it was enlarged to the 10x10 μm^2 SFM micrograph. The holes were produced artificially by the SFM tip upon tapping the surface of the liquid film.

In contrast to the uniform surface structure of the $HFPO_3$ polymer (Figure 1a), the HFPO-pentamer substitution caused a layer morphology as shown in Figure 1b. The perforated top layer was uniform in thickness, which was measured from a cross sectional profile to be 3.5 ± 0.5 nm. The observed morphology is typical for amphiphilic polymers, such as block copolymers and graft copolymers (24), and was also observed for the perfluorooctyl substituted PMA film in a smectic phase (17). Therefore, the formation of a well defined top layer in Figure 1b is consistent with the microphase separation into layers, comprised of alternating PMA and HFPO layers.

Film Stability. The same samples were imaged again after holding them at ambient conditions for 50 hours. This gentle treatment resulted in significant changes in the film structure as shown in Figure 2. The $HFPO_3$ polymer film ruptured forming circular holes, which were randomly distributed over the film surface (Figure 2a). One can hypothesize that dewetting occurred via nucleation of tiny holes, some of which could already be seen in the freshly prepared film in Figure 1a, followed by their growth and coalescence into larger holes in Figure 2a. This dewetting process stopped when the thickness of the dewetted layer reached 10 nm (profile in Fig 2a) . By scratching a small area at the bottom of the holes by the SFM tip in contact mode, it was revealed that the bottom was covered by a thin film of 1.6 ± 0.3 nm in thickness.

Unlike the HFPO-trimer substituted polymer, the originally flat film of the $HFPO_5$ substituted PMA coagulated into droplets with a defined contact angle of 10 ± 1 degrees. Also in this case, a thin film of the same material was found around the droplets. The film thickness was measured by SFM to be 2.5 ± 0.3 nm.

Thinner films of $HFPO_3$ and $HFPO_5$ substituted PMA were prepared by spincasting of more dilute solutions (c = 0.5 g/L) and investigated regarding their stability against dewetting. In both cases, SFM images showed a complete coverage of mica with a thin film of ca. 6 nm in thickness. However, typical dewetting patterns developed after 50 hours annealing at room temperature. The structures in Figure 3 appeared to be fully consistent with the micrographs of the thicker films, i.e the HFPO-trimer substituted PMA film ruptured into a perforated layer accompanied by thickening of the film. The film thickness of 12 nm, which was determined from the cross sectional profile in Figure 3a, is similar to the 10 nm thickness found upon dewetting of the thicker films. Still the substrate remained covered by a stable ultrathin film, which thickness was the

76

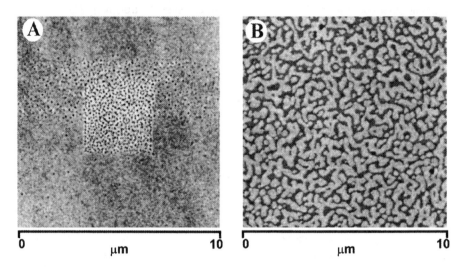

Figure 1. SFM height images of 12 nm thick films of $HFPO_3$ (A) and $HFPO_5$ (B) substituted PMA as prepared by spincasting on mica. The holes in the center of (A) emerged after scanning the 3x3 mm^2 area prior to the micrograph (A).

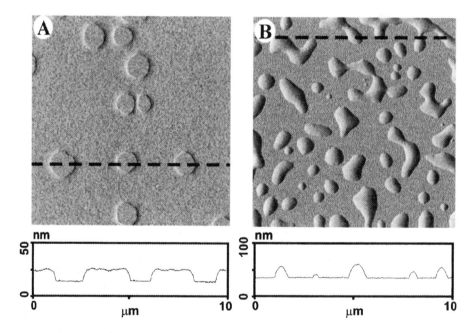

Figure 2. SFM amplitude images of the $HFPO_3$ (A) and $HFPO_5$ (B) substituted PMA films on mica after annealing at room temperature for 50 hours.

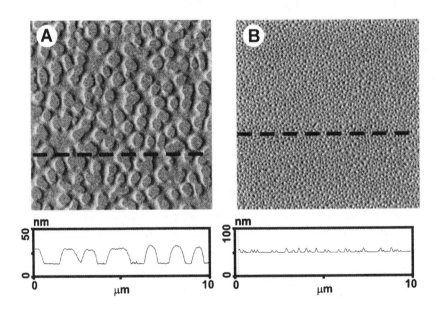

Figure 3. SFM images of the HFPO₃ (A) and HFPO₅ (B) substituted PMA films on mica after annealing at room temperature for 50 hours. Original thickness of the as spincast films was ca. 6 nm

same as in the case of the sample film prepared from the more concentrated solution, i.e., 1.6 ± 0.3 nm.

In contrast to the trimer sample, the HFPO-pentamer substituted PMA showed formation of droplets on a thin film of the same material (Figure 3b). The thickness of the stable film was measured by SFM to be 3 ± 0.5 nm.

Adhesion. The observation that the opened area at the bottom of the holes remained covered by a thin layer of the polymer was also verified by adhesion measurements between a SFM tip and the sample surface. Figure 4 shows three force-distance curves which were measured reproducibly by the same Si_3N_4 tip at the surface of the bare mica (Figure 4a), at the bottom of the holes in Figure 2a (Figure 4b) and the film in Figure 2b (Figure 4c). The area enclosed between the extension and retraction lines can be used to evaluate the adhesion work which was done upon breaking-off of the tip. The mica surface showed a high adhesion to the Si_3N_4 tip, affected by the capillary condensation, of water near the contact area (25). The adhesion curves in Figures 4b and c show unambiguously that the mica was exposed to air neither in the pentamer nor trimer samples. The diagram in Figure 4b is typical if the retraction of the tip involves stretching of a visco-elastic polymer which exhibits good adhesion to both mica and Si_3N_4 surface (26). The part of the material enclosed between the tip and the substrate could be stretched by more than 0.8 µm at a constant tip deflection, which means a constant force. In contrast to the above described diagrams, the film of $HFPO_5$ substituted PMA did not reveal measurable adhesion to the tip (Figure 4c).

A detailed discussion of the adhesion properties of the oligo(HFPO) substituted polymers is beyond the scope of this paper. The lower stickiness of the $HFPO_5$ substituted polymer could be due to the microphase separation between the fluorocarbon side chains and the hydrocarbon backbone affecting mechanical properties of the film (29).

Interpretation of the Wetting Behavior. In summary, the oligo(HFPO) substituted PMAs revealed fundamentally different wetting behavior depending on the number of HFPO units present in the side chains. The observed difference correlates with the microphase separation between the fluorocarbon and hydrocarbon blocks, which was detected only for $HFPO_5$ and $HFPO_4$ substituted polymer (Table 1). Small angle X-ray scattering did not give any evidence for regular packing of the microdomains in the bulk sample. However, in thin films, ordering could be favored by the preferential concentration of the $-CF_3$ groups at the free surface and the adsorption of the polar methacrylate groups to the mica surface (13). The specific interactions can result in an oriented monolayer, with a surface energy (γ_2) which is lower than that of the bulk material (γ_1). Although the exact value of the monolayer surface energy could not be measured by conventional surface techniques, macroscopic droplets, which emerged on the monolayer surface (Figure 4a), indicate somewhat different energies for the monolayer and the macroscopic phase. The phenomenon of droplet formation on an oriented layer of the same material is well known as autophobic wetting (27).

In contrast to the $HFPO_5$ sample, the $HFPO_3$ substituted PMA does not show a microphase separation which might favor surface segregation of the trifluoromethyl groups. Therefore, the surface structure of P($HFPO_3$-MA) films is expected to be less

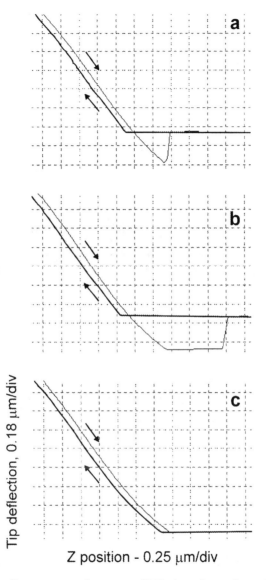

Figure 4. Force-distance curves between a Si_3N_4 tip and a surface of the following samples: bare mica (a); the bottom of the holes in Figure 2a.i.e., $HFPO_3$ methacrylate polymer (b); the film around the droplets in Figure 2b, i.e., $HFPO_5$ methacrylate polymer (c). All diagrams were measured reproducibly with the same tip.

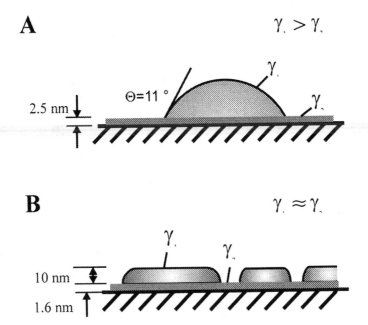

Figure 5. Schematic illustration of the pseudo partial wetting of mica by HFPO₅ (A) and HFPO₃ (B) substituted poly(methacrylate)s. While the droplets in (A) were ascribed to autophobic wetting (22), the thick film in (B) is probably caused by the chain elasticity.

uniform regarding chemical composition and orientation of the -CF$_3$ groups. The surface energies of such monolayer and of the bulk material could be very similar ($\gamma_1 \approx \gamma_2$), and therefore an autophobic wetting is not observed. Furthermore, the equilibrium thickness was shown to be independent of the surface coverage and tends to reach a thickness of ca. 10 nm. In addition to long range van der Waals forces, which can cause the film thickening over distances of more than 10 nm, it is more probable that the film breakage is caused by the entropic elasticity of the polymer chain (28). The 10 nm thickness fairly corresponds to the unperturbed dimensions of a hypothetical coil of a polyisoprene chain with a molecular weight of $1.8 \cdot 10^5$ g/mol (Table 1).

The difference in the wetting behavior of trimer and pentamer substituted PMAs can be also due to the difference in molecular weight. In this case, the film equilibration of the HFPO$_3$ substituted PMA, having a higher molecular weight, might take longer than for the lower molecular weight P(HFPO$_5$-MA). However, the kinetic arguments can be neglected as all three polymers are well above their glass transition temperatures. In addition, all film were tempered during several days and did not reveal any significan changes in the wetting behavior.

Conclusions

Polymethacrylates substituted with non-crystallizable oligo(HFPO) side chains were investigated regarding their wetting behavior on a mica substrate. Both polymers displayed pseudo partial wetting, as the originally flat film ruptured, however the opened area remained covered by a thin layer of the material. The dewetting phenomena in the HFPO$_5$ and HFPO$_3$ samples were different in origin and film structure. The pentamer substituted PMA displayed so-called autophobic wetting assisted by microphase separation and manifested by micrometer sized droplets on the oriented monolayer of the same material. In the case of the homogeneous sample of the HFPO-trimer substituted PMA, the dewetting process and the film thickness appeared to be controlled by the chain elasticity.

Acknowledgments. This work was financially supported by the Deutsche Forschungsgemeinschaft (project MO 682/2-1).

References

1. Kissa, E. Fluorinated Surfactants, Synthesis-Properties-Applications, Marcel Dekker, New York, 1994.
2. Zisman, W.A. in Contact Angle, Wettability and Adhesion; Gould, R.F., Ed.; ACS Adv. Chem. Ser., Washington, 1964, **1**, p.43
3. Jalbert, C.; Koberstein, J.T.; Hariharan, A.; Kumar, S.K. *Macromolecules* **1997**, *30*, 4481.
4. Jonson, R.E. Jr.; Dettre, R.H. *Polym. Prepr. (Am. Chem. Soc., Div. Polym. Chem)* **1987**, *28*, 48.
5. Doeff, M.M.; Lindner, E. *Macromolecules* **1989**, *22*, 2951.
6. Höpken, J.; Möller M. *Macromolecules* **1992**, *25*, 1461.
7. Hoyle, C.E.; Kang, D.; Jariwala, C.; Griffin, A.C. *Polymer* **1993**, *34*, 3070.
8. Katano, Y.; Tomono, H.; Nakajima, T. *Macromolecules* **1994**, *27*, 2342.

82

9. Kobayashi, H.; Owen, M.J. *Trends in Polymer Science* **1995**, *3*, 330.
10. Antonietti, M.; Henke, S.; Thünemann, A. *Adv. Mater.* **1996**, *8*, 41.
11. Höpken, J.; Sheiko S.S.; Czech, J.; Möller, M. *Polym. Prepr. (Am. Chem. Soc., Div. Polym. Chem)* **1992**, *33(1)*, 937.
12. Renz, W.; Warner, M. *Phys. Rev. Lett.* **1986**, *56*, 1268.
13. Fredrickson, G.H. *Macromolecules* **1987**, *20*, 2535.
14. Noirez, L.; Boeffel, C.; Daoud-Aladine, A. *Phys. Rev. Lett.* **1998**, *80*, 1453.
15. Brochard-Wyart, F.; Daillant, J. *Can. J. Phys.* **1990**, *68*, 1084.
16. Reiter, G. *Langmuir* **1993**, *9*, 1344.
17. Sheiko, S.S.; Lermann, E.; Möller, M. *Langmuir* **1997**, *12*, 4015.
18. Hayami, Y.; Findenegg G.H. *Langmuir* **1997**, *13*, 4865.
19. Gang, O.; Ocko, B.M.; Wu, X.Z.; Sirota, E.B.; Deutsch, M. *Phys. Rev. Lett.* **1998**, *80*, 1264.
20. Stoebe, T.; Mach, P.; Huang, C.C. *Phys. Rev. Lett.* **1994**, *73*,1384
21. Krupers, M.; Slangen, P.-J.; Möller, M. *Macromolecules* **1998**, *31*, 2552.
22. Slangen, P.-J. PhD Thesis, University of Twente, Enschede, The Netherlands 1998.
23. Sheiko, S.S.; Muzafarov, A.M.; Winkler, R.G.; Getmanova, E.V.; Eckert, G.; Reineker, P. *Langmuir* **1997**, *13*, 4172.
24. Russel, T.P. *Current Opinion in Coll. Interf. Sci.* **1996**, *1*, 107.
25. Weisenhorn, A.L.; Maiveld P.; Butt H.-J.; Hansma P.K. *Phys. Rev. B* **1992**, *45*, 11226.
26. Aimé, J. in *Scanning Probe Microscopy of Polymers*. Ratner, B.D.; Tsukruk, V.V. Eds.; ACS Symposium Series 694. American Chemical Society, Washington DC, 1998.
27. Zisman, W.A. *Industrial and Engineering Chemistry* **1963**, 55, 18.
28. Zhao, W.; Sokolov, J.; Rafailovich, M.; Kramer, E.J. *Phys. Rev. Lett.* **1993**, *70*, 1453.
29. de Crevoisier, G.; Fabre, P., Corpart, J.-M.; Leibler, L. *Science* **1999**, *285*, 1246

Chapter 7

Grand Unified Theory of Fluorocarbon Polymer Mobility on Disk Surfaces

C. Mathew Mate[1] and Robert S. Wilson[2]

[1]Almaden Research Center, IBM Research Division, 650 Harry Road, C3, San Jose, CA 95120
[2]Instructional Services Division, East Side Union High School District, San Jose, CA 95133

Often the effectiveness of a molecularly thin lubricant film depends on the polymer molecules moving to replenish those lost during sliding contacts. For spinning disks found in disk drives, many forces are present for driving the fluorocarbon lubricant molecules across the disk surfaces: diffusion, evaporation and redeposition, centrifugal acceleration, air shear, and interactions with the recording heads. We have developed experiments and analysis that enable us to study, in a unified manner, how these forces drive the movement of fluorocarbons on rotating disk surfaces. From these experiments, we are able to determine diffusion coefficients, effective viscosities, and the relative order of the driving mechanisms for several types of fluorocarbon lubricant systems.

If you open up the hard disk drive inside a personal computer, you will find a stack of disks as shown in Figure 1. Bits of information are stored in a thin layer of magnetic material deposited onto both sides of the disk. These bits of information are written and read by a recording head at the back of a slider that flies on a thin air film over each of the rotating disk surfaces. As the magnetic material is fairly soft, it is protected from occasional slider-disk contacts by a thin overcoat, usually amorphous carbon, and a molecularly thin film of lubricant. The preferred type of lubricant materials are perfluoropolyether polymers as they satisfy many of the disk drive industry's criteria (*1*) for these materials: liquid at the drive's operating temperatures, low surface tension, low volatility, and stable at high temperatures. Two examples of perfluoropolyethers typically used for disk drive applications are Fomblin ZDol-4000 and Demnum-SA-4000 (see Figure 2 for the chemical structures), which both have a linear polymer chain with alcohol end groups, a chain diameter of 7 Å, an average molecular weight of 4000 amu, corresponding to an end-to-end length of 120 Å. Since the average lubricant film thickness used in disk

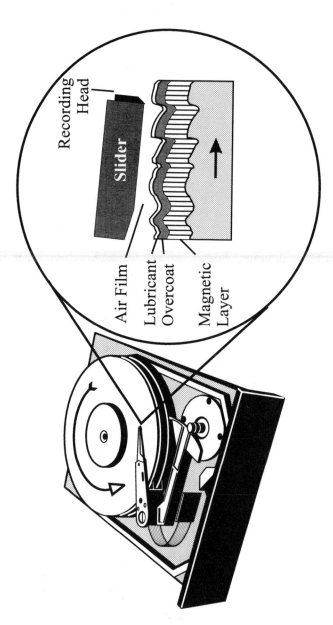

Figure 1. Schematic of the inside of a disk drive. Enlarged section shows the details of the slider-disk interface.

Fomblin Z-Dol: $HO-CH_2CF_2-[(OCF_2CF_2)_n-(OCF_2)_m]-OCF_2CH_2-OH$

Demnum SA: $CF_3-(OCF_2CF_2CF_2)_n-OCF_2CH_2-OH$

Figure 2. Chemical structure of perfluoropolyether lubricants.

drives typically ranges from 5 to 20 Å, the conformation of the lubricant molecules on the disk surface is mostly lying flat (2).

How these lubricant molecules move around on disk surfaces plays a key role in how well the lubricant film protects against slider contacts. During normal operation of the disk drives, a slider will occasionally strike the disk surface at high speeds; speeds greater than 10 m/s are common and can be as high as 50 m/s or 100 mph in the fastest spinning disk drives. During these contacts, the lubricant's job is to take as much of the impact and damage as possible so as to protect the underlying layers, while also providing the contacting interfaces with low shear resistance. Usually, the perfluoropolyether lubricants used in disk drives are terminated by reactive end groups, such as the alcohol end groups of ZDol and Demnum-SA, which anchor the lubricant molecules to the overcoat surface, so that molecules will be more difficult to squeeze out of the contact zone during the high speed impacts. This bonding of the lubricant molecules naturally greatly slows down the movement of these molecules over the surface. However, the slider-disk contacts are often so severe that a fraction of the lubricant molecules are detached from their anchoring points and ejected from the contact zone or, worst, are chemically degraded by the high shear forces or frictional heating generated by the contacting interfaces. After a slider impact has damaged the lubricant film in a particular spot, it is desirable for lubricant molecules in nearby areas to migrate quickly to the damaged area and replace the lubricant loss during the slider-disk impact. Consequently, two contradictory properties are desired from the lubricant molecules:
1) The ability to bond strongly to the disk surface and not be displaced during slider-disk contacts.
2) The ability to become unbonded and move quickly across the disk surfaces to replace those lubricant molecules lost in slider-disk contacts.

In this paper, we will discuss the various driving forces that move the lubricant molecules around on disk surfaces and describe an experimental procedure for separating the contributions of the various driving forces responsible for moving the lubricant on rotating disk surfaces.

Driving Forces for Moving Lubricants in Disk Drives

Inside a disk drive, as depicted in Figure 1, many forces are responsible for moving the lubricant molecules around on the disk surfaces. These forces can be divided in three general categories: 1) molecular forces arising from the interactions of lubricant molecules with the underlying disk surface and with each other, 2) forces generated by the rotating disk, and 3) the forces generated by a slider flying over the disk surface and occasionally contacting the surface. Each of these general categories consist of several driving mechanisms:
1. **Molecular forces**
   I. **Diffusion.** When a gradient in lubricant thickness ∇h exist on a disk surface, lubricant molecules will diffuse toward the thinner areas. The flow q of molecules across the surface from diffusion is described by Fick's first law:
$$q = D \nabla h$$
   where D is the diffusion coefficient.

II. **Evaporation.** As the temperature is increased, more molecules will have enough thermal energy to evaporate from the surface. Fortunately for the lubricants shown in Fig. 2, the rate of evaporation remains small at most disk drive operating temperatures (20-60°C), but can become significant at temperatures above this range. When a molecule evaporates from the disk surface, it doesn't travel far, but instead undergoes Brownian motion in the air immediately above the disk surface and eventually may be redeposited at a new location on the disk surface. Air flow across the disk surface will of course help to carry the lubricant molecules away from the location from which they originally evaporated.

2. **Forces from the rotating disk**

   I. **Centrifugal.** The centrifugal force experienced by a molecule on a rotating disk is

$$F = m_{mol}\omega^2 r$$

where m_{mol} is the mass of the lubricant molecule, ω is 2π times the disk rotation frequency, and r is the distance of the molecule from the center of the disk. *Example calculation of magnitude:* For a lubricant molecule like Fomblin ZDol-4000 ($m_{mol} = 6.5 \times 10^{-21}$ gm) sitting at the edge of a 95 mm diameter disk spinning at 10,000 rpm the centrifugal force is $\underline{3 \times 10^{-19}}$ N/molecule, which corresponds to an acceleration of 5300 g.

   II. **Air Shear.** As a disk rotates, a thin film of air next to the disk surface is carried along with the disk surface. The air film has a thickness $\delta = \sqrt{\eta_{air}/\rho_{air}\omega}$ where η_{air} and ρ_{air} are, respectively, the air viscosity and density (*3*). The centrifugal forces acting on this rotating air film slowly drive the air molecules in the film towards the outer edge of the disk. The radial component of air velocity towards the OD of the rotating disk surface generates a radial component of the air shear force or stress on the lubricant film that pushes the lubricant molecules towards the outer edge of the disk, which can be several orders of magnitude greater than the direct centrifugal force acting on the lubricant film (*4*). For a disk surface with no other surfaces nearby it, like the top surface of the disk stack shown in Figure 1, the radial component of the air shear force acting on an individual molecule in the lubricant film on a rotating disk without any nearby surfaces can be expressed as

$$F_r = A_{mol}\tau_r$$

$$= A_{mol}\rho_{air}r\omega\sqrt{\eta_{air}/\rho_{air}\omega}$$

where A_{mol} is the effective area of a lubricant molecule in the lubricant film and τ_r is the radial component of the stress from the air shear acting on the top of the lubricant film. *Example calculation of magnitude:* For a Fomblin ZDol-4000 lubricant molecule in a 1 nm thick film near the OD of a 95 mm diameter disk rotating at 10,000 rpm, $A_{mol} = (3.4 \text{ nm}^3 / 1 \text{ nm}) = 3.4 \text{ nm}^2$, then $\underline{F = 2.6 \times 10^{-17} \text{ N/molecule.}}$

3. Slider forces

I. **Slider generated air pressure gradients.** For sliders used in disk drives, the surface that faces the disk is especially designed to be an air bearing surface that compresses air as it enters underneath the front edge of the slider. The high air pressures (as much as 10 atmospheres) generated between the slider's air bearing surface and the disk provide the lift that counteracts the applied load from the slider suspension, enabling the recording head on the slider to fly at a nearly constant height over the disk surface. These high pressures are also exerted on the top of the lubricant film and will try to push the lubricant molecules out from underneath the slider body. The force acting on the individual lubricant molecules comes from ∇p, the gradient of air pressure laterally across the disk surface,

$$F = V_{mol} \nabla p$$

where V_{mol} is the volume of a lubricant molecule. *Example calculation of magnitude:* Assume that the pressure changes by 10 atm (1 MPa) over a 100 μm distance across the disk surface and that $V_{mol} = 3.4$ nm^3. Then, F = 3.4x10^{-17} N/molecule.

II. **Slider generated air shear.** The pressurized air film between the slider and the disk is also being sheared at a high rate by the rotating disk underneath the slider. The shearing of this air film generates opposite acting shear stresses on the slider and disk surfaces that push the lubricant molecules on the disk surface in the direction of the slider motion. The slider air shear force can be expressed by

$$F = \frac{\eta_{air} V A_{mol}}{D}$$

where V is the velocity of the disk relative to the slider and D is the separation distance or "fly height" of the slider above the disk surface. *Example calculation of magnitude:* If $D = 40$ nm and $V = 50$ m/s, then F = 7.5x10^{-14} N/molecule.

III. **Slider contact shear.** When the slider physically contacts the disk surface, gigantic shear rates and shear stresses are generated by the high impact speeds that will not only displace unbonded lubricant molecules from the contact zone, but can potentially break internal chemical bonds shredding those molecules remaining in the zone. These interactions truly represent the "Godzilla" of the forces experienced by the lubricant molecules. The contact shear forces can be approximated by

$$F = \frac{\eta_{lub} V A_{mol}}{h}$$

where η_{lub} is the effective viscosity of the lubricant in the contact zone and h is the thickness of the lubricant film being sheared. *Example calculation of magnitude:* If $\eta_{lub}=0.2$ Pa·s, $h = 10$ Å, $V = 50$ m/s, $A_{mol} = 3.4$ nm^2, then F = 3x10^{-8} N/molecule, a force sufficient enough to rupture the internal bonds of the lubricant molecules. This estimated contact shear force is also eleven orders of magnitude higher than the estimate for the centrifugal force acting on the film. Of course, in today's disk drives contact forces act only for very brief moments while the centrifugal force is always present.

Previous Experimental Work

Several earlier publications have studied the movement of perfluoropolyether lubricants on surfaces. Diffusion is generally studied by depleting an area of lubricant then measuring the diffusion of lubricant back into that area (5,6). Several publications have measured the spin-off of lubricants from rotating disk surfaces (7-9). The evaporation-redeposition mechanism for several non-perfluoropolyether molecular films have also been reported (10,11).

Measurements of Lubricant Movement on Rotating Disk Surfaces

In this paper we report a experimental methodology for measuring the various contributions to the movement of lubricant on rotating disk surfaces without sliders. The basic idea is to deposit lubricant onto a disk surface so that it has more lubricant at the ID than at the OD, then to monitor the movement of lubricant toward the OD when subjected to different conditions that emphasize the different driving forces: diffusion, evaporation, centrifugal, and air shear.

Experimental Method. Figure 3 shows schematically how the lubricants are first deposited uniformly onto the disk surface then removed from an annulus at the OD. For the initial deposition step, the disk is submerged in a solution of the lubricant is dissolved in a volatile solvent (perfluorohexane) and then withdrawn at a constant speed. As the disk is withdrawn, it pulls up a thin film of solution, the thickness of which increases with pull speed. The solvent quickly evaporates from this thin film leaving behind a uniform thickness of lubricant that depends on the initial concentration of lubricant in the solution and the withdrawal speed from the solution (12). Next, the lubricant at the outer diameters is removed by rotating the disk while it is partially submerged in a beaker containing pure solvent. This procedure creates a sharp step in the radial profile of lubricant thickness on both sides of the disk. We determined the thickness profile by using grazing angle Fourier Transform Infrared (FTIR) Spectroscopy to measure the intensity of the asymmetric CF stretching mode at 1285 cm^{-1}, which was calibrated against the lubricant thickness using a combination of ellipsometry and X-ray reflectivity (13). For these experiments, we used an FTIR beam with a 3 mm diameter circular spot size on the disk surface.

ID-to-OD step profiles in lubricant thickness with the same type of lubricant are created on two disks. One of these disks is placed, not spinning, in an oven at the experimental temperature; lubricant on this disk only moves from diffusion and evaporation. The other disk is placed either at the top or bottom of a disk stack as shown in Figure 4, and the stack placed in the oven. Rotating the disks in the disk stack with only a small gap (0.25 mm) between them, forces all the air between the disks to rotate with the disks so that no air shear is generated on these inner surfaces. Two experiments with different types of lubricant can be run per rotating spindle. A blank, unlubricated disk is used to separate the top and bottom disks so that lubricant evaporating from one of the inner experimental surfaces does not contaminate the other inner surface. The evolution of the lubricant thickness profile with time is

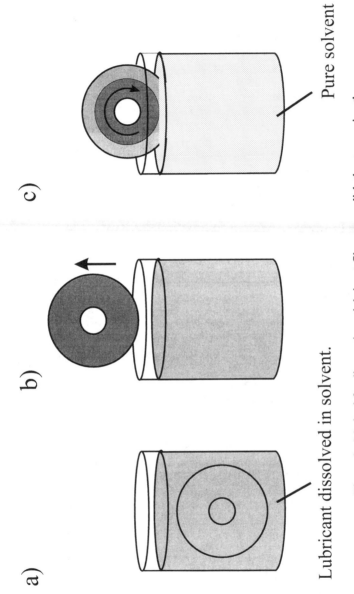

a)

b)

c)

Lubricant dissolved in solvent.

Pure solvent

Figure 3. Method for dip coating a lubricant film onto a disk then removing the film from the outer edge.

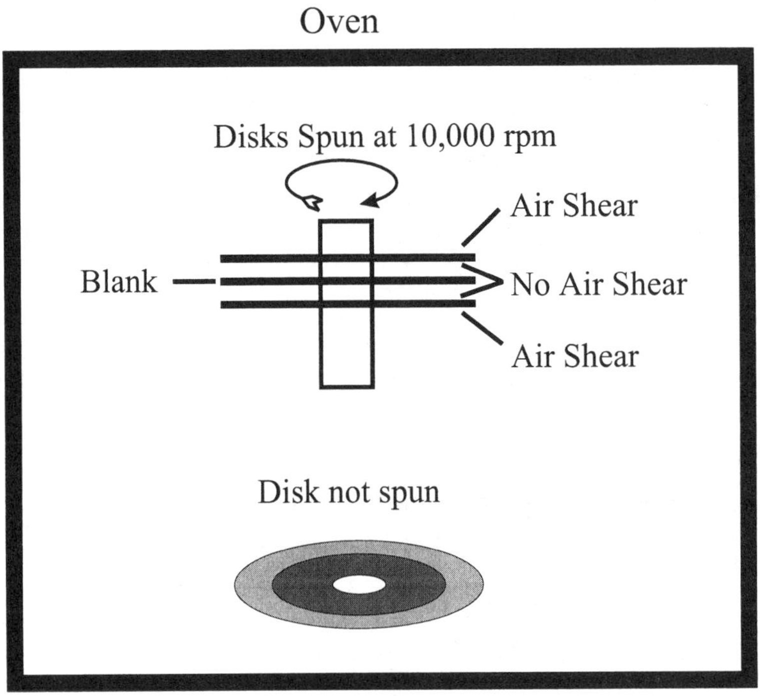

Figure 4. For polymer lubricant mobility studies, disks with stepped lubricant profiles are either placed at the top or bottom of a spinning disk stack or not spinning on the floor of the oven.

monitored by removing the disks occasionally from the disk stack and oven and remeasuring the profile at the same angular location along the disk surface.

Results. Figure 5 shows experimental results obtained for three disk surfaces with profiles prepared with Demnum-SA-4000 and exposed to the different driving conditions. The thin film disks used consisted of 95 mm diameter AlMg substrates with a NiP overlayer onto which a cobalt alloy magnetic layer and an amorphous hydrogenated carbon overcoat were sputter deposited. The resulting disk surface is fairly smooth, 1.5 nm rms roughness over a 1 μm^2 area as determined by AFM. A fairly high temperature of 100°C was used to accelerate lubricant movement. From the profiles for the disk that does not spin (Figure 5a), it is readily apparent that, in addition to diffusion of the Demnum-SA-4000 lubricant toward outer radii, a significant amount of lubricant is being lost to evaporation at this elevated temperature. Figure 5b shows the profiles for the inner surface spinning at 10,000 rpm. For this surface the movement of the step edge toward outer radii is only slightly faster from the addition of the centrifugal force to the diffusion driving force. However, the centrifugal force noticeably speeds up the thinning of the lubricant film at the inner radii. In particular, the lubricant becomes more depleted at the innermost radius as lubricant that is driven outward by the centrifugal force cannot be replace due to the presence of the hole at the center of the disk. Figure 5c shows the profiles for the surface exposed to air shear, where it is readily apparent the air shear is the dominant driving force, quickly depleting Demnum-SA-4000 from the inner radii by driving it to the outer radii.

Effect of End Groups and Molecular Weight. Figure 6 shows the profiles for three non-spinning disks with different types of lubricants at 65°C, where the lubricant moves mainly by diffusion. At this lower temperature, all three lubricants have very little evaporation. The lubricants were chosen to show the effect that end groups and molecular weight have on diffusion (all three have the same type of polymer backbone). An estimate for the diffusion coefficient D can be obtain using the relation $D = L^2/t$ where L is the distance that front (measured at some thickness such as 10 Å) has moved in a time t. Applying this relation to the data in Figure 6, we find that $D = 8 \times 10^{-11}$ m^2/s for Demnum-S65, $D = 2 \times 10^{-11}$ m^2/s for Demnum-SA-4000, and $D = 0.4 \times 10^{-11}$ m^2/s for Demnum-SA-5500. From these values for the diffusion coefficients, we see that Demnum-S65, which has neutral -CF$_3$ end groups at both ends of the polymer chain and a molecular weight of 4500, diffuses the fastest (Figure 6a) and that the rate of diffusion slows significantly when an alcohol end group is added for Demnum-SA-4000. The rate of diffusion slows even more when the molecular weight is increased to 5500 amu (Demnum-SA-5500, Figure 6c). The lower diffusion coefficient with addition of the alcohol end group presumably comes from the strong interaction between end group and the disk surface causing the end group to act like an anchor. This trend in diffusion coefficient is very similar to the trend observed in effective viscosities observed for these lubricant films under conditions where centrifugal forces are driving the molecules, as described in one of our recent papers (*1*).

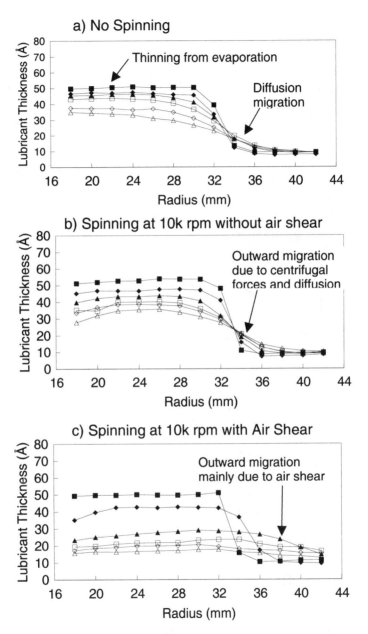

Figure 5. Profiles for disks, lubricated with Demnum-SA-4000, under different driving conditions, all at 100°C. The profiles correspond to times 0, 3, 24, 48, 69, and 140 hours.

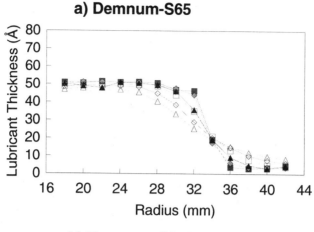

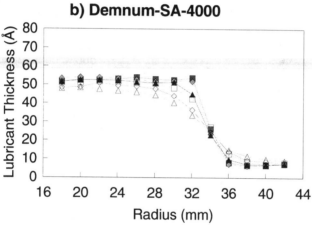

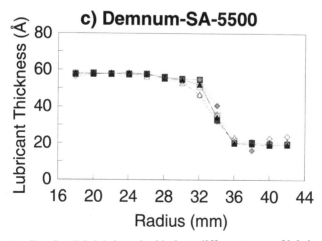

Figure 6. Profiles for disk lubricated with three different types of lubricant, not spinning at 65°C. Profiles correspond to times 0, 3, 20, 45, 92, 162 hours.

95

Conclusions

Many types of forces can move lubricants around on disk surfaces in disk drives: molecular forces such as diffusion and evaporation; rotation forces such as centrifugal and air shear; and slider forces such as air pressure and shear and contact pressure and shear. Here we have shown an experimental procedure for separating out the different contributions for driving lubricant molecules on rotating disks without sliders. We find that air shear forces are the dominant driving force, followed by diffusion, then centrifugal forces. Lubricant losses due to evaporation can be significant at higher temperatures. The additions of an alcohol end group and increasing the molecular weight act to slow down the rate of lubricant movement on the disk surfaces.

Acknowledgments

R. S. Wilson would like to acknowledge funding from the NSF/San Jose State University Analytical and Surface Chemistry of Materials Research Training Program.

Literature Cited

1. Mate, C. M. *Tribology Letters* **1998**, *4*, pp 119-123.

2. Mate, C. M.; Novotny, V. J. *J. Chem. Phys.* **1991**, *94*, pp. 8420-8427.

3. Schlichting, H. *Boundary Layer Theory*, 7th Edition; McGraw-Hill: New York, NY, 1979; pp. 102-107.

4. Middleman, S *J. Appl. Phys.* **1987**, *62*, pp. 2530-2532.

5. Novotny, V. J. *J. Chem. Phys.* **1990**, *92*, pp. 3189-3196.

6. O'Connor, T. M.; Jhon, M.S.; Bauer, C.L.; Min, B.G.; Yoon, D.Y.; Karis, T.E. *Tribology Letters* **1995**, *1*, pp. 219-223. Min, B.G.; Choi, J.W.; Brown, H.R.; Yoon, D.Y.; O'Connor, T.M.; Jhon, M.S. *Tribology Letters* **1995**, *1*, pp. 225-232.

7. Yanagisawa, M. *J. Appl. Phys.* **1987**, *61*, pp. 1034.

8. McConnell, W. H. *J. Appl. Phys.* **1988**, *64*, pp. 2232.

9. Forcada, M. L.; Mate, C. M. *J. Colloid Interface Sci.* **1993**, *160*, pp. 218-225.

10. Novotny, V. J.; Marmur, A. *J. Colloid Interface Sci.* **1991**, *145*, pp. 355-361.

11. Forcada, M. L.; Mate, C. M. *Nature* **1993**, *363*, pp. 527-529.

12. Gao, C; Lee, Y. C; Russk, M. *IEEE Trans. Magn.* **1995**, *31*, pp. 2982-2984.

13. Toney, M. F.; Mate, C. M.; Pocker, D. *IEEE Trans. Magn.* **1998**, *34*, pp. 1774-1776.

Chapter 8

Fluorinated Silicone Coatings with Controlled Surface Chemistry and Microstructure

Mattias Berglin[1], Erika Johnston[2,3], Kenneth Wynne[2,4], and Paul Gatenholm[1]

[1]Department of Polymer Technology, Chalmers University of Technology, S-412 96 Göteborg, Sweden
[2]Materials Chemistry Branch, Naval Research Laboratory, Code 6120, 4555 Overlook Avenue, SW, Washington, DC 20375
[3]Chemistry Department, Virginia Polytechnic Institute and State University, Blacksburg, VA 24061
[4]Office of Naval Research, Boston Towers, Arlington, VA

We have demonstrated that the surface chemistry and surface morphology of a sol-gel system composed of poly(dimethylsiloxane) (PDMS) and (tridecafluoro-1,1,2,2-tetrahydrooctyl) triethoxysilane (FTEOS) cross-linker can be controlled by varying the catalyst concentration, molecular weight of PDMS and humidity during curing. ESCA analysis revealed that increasing the catalyst concentration and humidity during curing resulted in up to eight times enrichment of fluorine at the surface. The variation in surface topography from smooth to micro-structured was visualized with AFM. The structures formed mainly with low molecular weight PDMS and at medium to high catalyst concentrations were found to be in the range important for directing the biological response.

The interactions between living organisms or living tissue and a foreign material are governed by the physical and chemical properties of the material. The surface properties control reactions such as wetting and adhesion occurring at the material/organism interface. The surface properties may act independently or in synergy to produce biological responses to foreign materials. It has been shown, for example, that surface properties can affect the cell response e.g. cell growth. The surface-directed cell growth has been shown to occur on both topographically modified surfaces [1-2] and on surfaces having patterned chemical properties, such as alternating rows of adhesive/non-adhesive properties [3-4].

Phase separation of polymers has recently been pointed out as an attractive technique for preparing structures of controlled size and morphology [5]. Recent studies have focused on structure formation on coatings with a deformable polymer-air interface [6-7]. In this type of experiment, surface tension has a strong influence on the boundary shape where an interfacial free energy minimization process is responsible for the surface structures formed. The built-in flexibility of this interface can cause the film surface to form unique patterns depending on the underlying phase separation process [8]. Another way to prepare structures is by the formation of self-assembled aggregates or micelles occurring in certain block copolymers [9]. In this type of system, conformation is dependent on rigid and soft segments of the copolymer. Diblock copolymers have been shown to form cubic arrays, spheres, hexagonal arrays of cylinders bicontinuous cubic phases or lamellae, depending on the relative block lengths [10].

Surface segregation of fluorine-ended monomers and fluorinated block copolymers has been used to engineer the surface chemistry [11-12]. The use of fluorinated polymers is attractive in minimally adhesive applications such as stain resistant fabrics, medical implants that resist cell-adhesion, self-lubricating surfaces for magnetic recording media and non-toxic fouling release coatings for ship hulls. The surface enrichment of fluorine makes it possible to prepare hydrophobic surfaces despite a low bulk composition of fluorine.

Sol-gel chemistry based on poly(dimethylsiloxane) has been found very useful for preparing coatings that exhibit minimal adhesion of marine organisms [13-14]. Low bioadhesion combined with surface structures that can give rise to a controlled biological response may be one way to prepare non-toxic surfaces that will reduce the attachment of marine macrofoulers such as mussels, barnacles, hydroids and algae. It has recently been shown that it is possible to alter the chemical composition and morphology of surfaces of PDMS networks prepared by sol-gel chemistry by introducing various amounts of fluoralkyl triethoxysilane cross-linker [15]. In this study, the surface chemistry and surface morphology were controlled by varying the catalyst concentration, molecular weight of PDMS and humidity during curing.

Experimental

Materials: Silanol terminated polydimethyl siloxane (PDMS) (4.2 kDA, cat# DMS-S21, 26kDa, cat# DMS-S31; 49 kDa, cat# DMS-S35) and (tridecafluoro-1,1,2,2-tetrahydrooctyl) triethoxysilane (FTEOS) (cat #SIT8175, CAS[51851-37-7]) were obtained from Gelest (Tullytown, USA). The catalyst, Dibutyltin diacetate (DBTDA), was purchased from Aldrich, Sweden (cat# 29,089-0, CAS[1067-33-0]).

Film formation: Dibutyltin diacetate (DBTDA) catalyst was added dropwise with a syringe to a polymer and alkoxysilane solution under stirring. After 30 minutes, the solution was degassed and 1.2 +/- 0.3 mm thick films were spin-casted in polystyrene petri dishes. The films were placed in controlled humidity for network formation. The ethoxy/hydroxyl stoichiometric ratio was kept constant at six for all conditions. Equations 1 and 2 show the reactions responsible for network formation by FTEOS [16]. The equations below assume that ethoxy groups on the cross-linker hydrolyze in the presence of moisture to release ethanol and that the silanol groups formed by hydrolysis condense with the silanol-terminated PDMS. FTEOS may also self-condense, which is displayed in equation 3 [16]. Self-condensation reactions involve the net incorporation of one-half mole of water for each mole of ethanol released.

$$\text{CF}_3\text{-(CF}_2)_5\text{-(CH}_2)_2\text{-Si(OEt)}_3 + 3\text{H}_2\text{O} \xrightarrow{\text{DBTDA}} \text{CF}_3\text{-(CF}_2)_5\text{-(CH}_2)_2\text{-Si(OH)}_3 + 3\text{EtOH}$$ (1)

$$\text{CF}_3\text{-(CF}_2)_5\text{-(CH}_2)_2\text{-Si(OH)}_3 + 3/2 \text{ HO-PDMS-OH} \xrightarrow{\text{DBTDA}} \text{CF}_3\text{-(CF}_2)_5\text{-(CH}_2)_2\text{-Si-(O-PDMS-OH)}_3 + 3\text{H}_2\text{O}$$ (2)

$$\text{CF}_3\text{-(CF}_2)_5\text{-(CH}_2)_2\text{-Si(OEt)}_3 + 3/2 \text{ H}_2\text{O} \xrightarrow{\text{DBTDA}} \text{CF}_3\text{-(CF}_2)_5\text{-(CH}_2)_2\text{-Si-O}_{3/2} + 3\text{EtOH}$$ (3)

Experimental design: Three variables, catalyst concentration (0.05–5 w%), molecular weight (4200-49000 g×mol^{-1}) and humidity during curing (0-100% relative humidity) were investigated. A factorial experimental design was used to study the effects of the variables on the responses. In an experimental design, large effects on the response (the behavior of the response in the experimental region) can be investigated, as can synergy effects. A 2^3 factorial design led to eight experimental conditions. The experimental conditions are summarized in Table I and presented graphically in Figure 1. The centerpoint experiments were repeated three times whereas the experiments at the corners of the experimental design were not repeated. We assumed the experimental variation in the centerpoints to be comparable with the experimental variation at the corners of the design. With this assumption we considered it sufficient with one experiment per experimental condition. The experimental conditions under which surface microstructures were formed in a previous study [15] were chosen for the centerpoint of the factorial design. To increase the explanatory degree of the model it was necessary to carry out six extra experiments (sample ID 9-14). The design was created in Modde version 4.0 (Umetri AB, Sweden) and evaluated with multiple linear regression (MLR) as the regression method. The parameters used in this paper, R^2 (the fraction of variation of the responses that can be explained by the model) and Q^2 (the fraction of variation of the responses that can be predicted by the model), are given by:

$$R^2 = SS_{reg}/SS$$ (4)

$$Q^2 = 1 - PRESS/SS$$ (5)

where SS_{reg} is the sum of squares of Y, corrected for the mean, explained by the model. SS is the total sum of squares of Y corrected for the mean, and $PRESS$ is the prediction residuals sum of squares. R^2 and Q^2 are used as indicative criteria of the model fit. To further check the experimental data and prediction capability of the model, two extra experiments were carried out. These two experiments (sample ID 15-16) were used to compare the predicted results with analyzed.

The responses we studied were the enrichment of fluorine, investigated with electron spectroscopy for chemical analysis (ESCA), and the surface morphology, using atomic force microscopy (AFM). The enrichment of fluorine can be defined as:

$$Enrichment = \frac{Atomic\ surface\ concentration\ of\ fluorine}{Atomic\ bulk\ composition\ of\ fluorine\ (theoretically)} \quad (6)$$

Table I: Experimental conditions

Sample ID	W% Catalyst Concentration	Relative Humidity (%)	Molecular weight (kD)	Bulk composition of fluorine (%)
1	0.05	0	4.2	15.8
2	5	0	4.2	15.4
3	0.05	100	4.2	15.8
4	5	100	4.2	15.4
5	0.05	0	49.0	2.0
6	5	0	49.0	1.9
7	0.05	100	49.0	2.0
8	5	100	49.0	1.9
9	0.05	53	26.0	3.9
10	5	53	26.0	3.7
11	0.5	0	26.0	3.8
12	0.5	100	26.0	3.8
13	0.5	53	4.2	15.7
14	0.5	53	49.0	2.0
Centerpoint (x3)	0.5	53	26.0	3.8
15	0.05	100	26.0	3.8
16	5	100	26.0	3.8

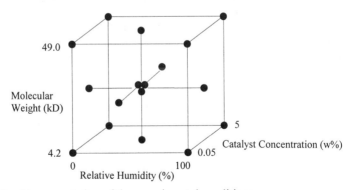

Figure 1. Graphic presentation of the experimental conditions.

ESCA: Electron spectroscopy for chemical analysis (ESCA) was carried out using a Perkin Elmer PHI 5500 instrument. The X-ray source was a Mg K_α anode, and all measurements were performed at a 45-degree take-off angle. Binding energy shifts were resolved using the peak for hydrocarbon at 285.0 eV as a reference.

Tapping mode AFM: The surface morphology was examined with a Digital Instruments NanoScope III atomic force microscope fitted with a NanoScope IIIA controller and a Dimension 3000 large sample type G scanner. The films were analyzed in tapping mode with standard silicon tips. RMS amplitude, setpoint and frequency used is displayed in each figure.

Controlled humidity: Humidity was controlled in an airtight glass container containing saturated salt solutions according to ASTM standard E 104–85. The experimental condition of 0% relative humidity was obtained in an airtight glass container filled with silica gel.

Results and Discussion

Film formation: The appearance of the fully cured films ranged from transparent and shiny (samples ID 2-6, 8-12, 14 and centerpoint) to cloudy and dull (samples ID 1,7 and 13). The curing time (the time until the surface of the film felt non-tacky) ranged from hours for the films cured at high humidities to a few days for the films cured at low humidities. Sample ID 6, 8 and 10 never cured to totally non-sticky films. Even after four weeks in controlled humidity the films were still sticky, an evidence of still uncross-linked PDMS chains. The sticky samples all contained high amounts of catalyst and medium to high molecular weight PDMS polymer (Table 1). The kinetics of self-condensation of FTEOS may explain this phenomenon. If FTEOS self-condenses to a high extent, no free hydroxyl groups will be available for network formation. Black et al. [19] obtained chain extensions instead of self-condensation of tetraethyl orthosilicate (TEOS) at low catalyst concentrations. When the catalyst concentrations were increased the self-condensation of TEOS gave raise to micron-size particles in the bulk, which was revealed with small angle neutron scattering (SANS). Similarly, the self-condensation of FTEOS at higher catalyst concentrations will stop the formation of a polymer network. No analysis of sample ID 10 was possible owing to the tackiness. Sample ID 6 and 8 were possible to analyze. This may be due to the higher molecular weight used in the samples. In comparison, sample ID 2 were low molecular weight PDMS polymer was used a rigid network with no tackiness was formed.

The first step in the curing reaction is the hydrolysis of FTEOS, at which ethanol is released. This is followed by a condensation reaction and the formation of a covalent bond between FTEOS and PDMS. Water is required in the hydrolysis step, but is formed at the last condensation step. The net usage of water is therefore zero, and a low moisture level is sufficient to drive the reaction. The experiments carried out at 0% relative humidity were of course not totally free of water moisture but under the detection limit for the electronic hygrometer used to measure the humidity.

Surface Chemical Composition: Changing the variables in the experimental design resulted in a large variation in the surface elemental composition, as revealed by ESCA. Table II presents the fluorine concentration at the surface for the different films. The concentration of fluorine at the surface ranged from roughly bulk composition (sample ID 1) to nearly eight times bulk composition (sample ID 14). The presence of the CF_3, CF_2 and CH_2 peaks from the cross-linker together with the CH_3 peak from the poly(dimethylsiloxane) can be seen when the carbon peak is resolved for sample ID 12, Figure 2. A ratio of 5:2:1 is obtained when analyzing the size of the CF_2, CH_2 and CF_3 peaks, which is in agreement with the composition of FTEOS.

Table II. Atomic concentration of fluorine

Sample ID	Fluorine content measured at T.O.A. 45° (%)	Enrichment (measured/bulk composition)
1	14.6	0.92
2	43.5	2.82
3	44.4	2.82
4	35.4	2.29
5	9.88	4.94
6	11.7	6.05
7	13.3	6.65
8	12.3	6.37
9	12.5	3.18
11	11.7	3.06
12	27.0	7.04
13	24.7	1.57
14	15.6	7.84
Centerpoint (mean of 3 films)	21.0	5.48

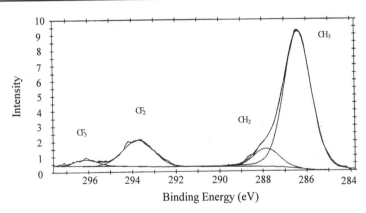

Figure 2. ESCA, C1s peaks collected at take-off angle of 45° for sample ID 12

The importance of humidity and catalyst concentration was clearly seen when fitting a model to the experimental data and using it to predict the enrichment over the entire experimental region, Figure 3. The figure should be read as a topographical map where the enrichment contours are displayed in the z-direction (going outwards from the graph). Evaluation of the statistical model explained about 88% of the experimental data (R^2). The predictability (cross-validation, Q^2) of the statistical model was 71%. The catalyst concentration was transformed to 10log in order to increase the model fit. ANOVA calculation showed no lack of fit, which means that the model is significant at the 5% confidence level. The normal probability plot of the residuals (the mismatch between the observed and modeled values) showed no outliers.

The driving force for the enrichment of fluorine towards the surface is the minimization of the interfacial tension at the polymer/air interface. This surface segregation of fluorine has been demonstrated, as mentioned above, by other authors [11-12]. In this study, the fluorine concentration is slightly higher at the surface when the curing humidity is increased, Figure 3. This appears to stand in contrast to the minimization of the interfacial surface tension responsible for the surface migration. However, this observation may be explained by a fast hydrolysis of FTEOS with a corresponding self-condensation of FTEOS to higher molecular weight aggregates. This increases the tendency for the system to phase separate according to the Flory-Huggins theory of polymer miscibility, resulting in a increased enrichment of fluorine at the surface. In a sol-gel system consisting of PDMS and tetraethyl orthosilicate (TEOS), the degree of self-condensation has been observed to increase with increasing water content during curing [18]. FTEOS will probably behave like TEOS, with self-condensation and increased phase separation tendencies at higher humidities.

Increasing the catalyst concentration from 0.05 % (w/w) to 0.5 % (w/w) increased the surface concentration of fluorine. Increasing the catalyst concentration further to 5 % (w/w) reduced the fluorine enrichment slightly. The thermodynamics of the phase separation and chemical kinetics of the cross-linking reaction are probably the reasons for the different results achieved. Black et al. [19] obtained chain extensions instead of self-condensation of tetraethyl orthosilicate (TEOS) at low catalyst concentrations. When the catalyst concentrations were increased the self-condensation of TEOS gave raise to micron-size particles in the bulk, which was revealed with small angle neutron scattering (SANS). Similarly, the self-condensation of FTEOS at higher catalyst concentrations with an increased tendency toward phase separation may increase the concentration of fluorine at the surface.

The chemical kinetics of the cross-linking reaction and the thermodynamics of the phase separation must be studied in more detail before the enrichment mechanisms can be fully explained. The chemical kinetics of the reaction, with the possibility of self-condensation of the FTEOS, may also play a role in the surface enrichment of fluorine.

It should be noted that Figure 3 is based on a statistical model and, in order to control its accuracy, two extra films were prepared. The two new films were analyzed, and the enrichment of fluorine was compared with predictions from the model. The two new compositions together with the measured and predicted enrichment are shown in Table III, where the 95% lower and upper confidence intervals for the predicted data are also given. On the basis of these two new experiments, it can be concluded that the model can be used for a rough estimate of surface enrichment but more experimental work must be performed before a more accurate enrichment can be predicted.

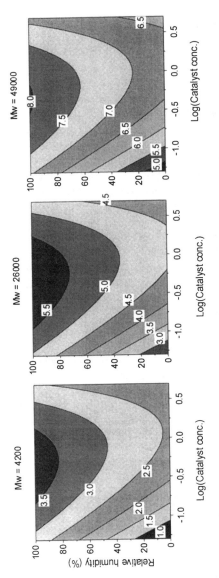

Figure 3. Contour plots based on the statistical model, showing the enrichment of fluorine at the surface.

104

Table III. Control of model accuracy

Sample ID	W% Catalyst	Rh (%)	Mw (kD)	Enrichment of fluorine (measured)	Enrichment of fluorine (predicted)	Predicted enricment Lower 95%	Prediced enrichment Upper 95%
15	0.05	100	26.0	5.68	4.85	3.54	6.16
16	5	100	26.0	5.11	4.64	3.26	6.02

Surface Morphology: The surface morphology studied with AFM (investigated in samples ID 1-8, 11-12 and centerpoint) showed a distinct variation. The surfaces ranged from smooth (Figure 4) to craterous to islands of regular structure and size (Figure 5). The average diameter of the structures ranged from about 0.5 μm, sample ID 5, to over 6.5 μm, sample ID 2. The height of the islands ranged from about 0.1 μm, sample ID 5, up to a maximum of 0.5 μm, found on sample ID 2. The structures were present over the entire film surface but the distances between the islands differed among the films. The average diameter of the structures on the films were determined with AFM and is shown in table IV. The strong influence of catalyst concentration is obvious when studying data from sample ID 1-4. On sample ID 7, the structures were changed to small "holes" spread out over the surface instead of the raised structures found on the other films. An average "hole" diameter of 1.0 μm was obtained when the top of the holes was measured. Table IV gives the average domain size of film ID 7 as zero since the structures were so vastly different than the other films. The aspect ratios (height H to diameter D, $\alpha = H/D$) of the structures were all in the range of 0.06 to 0.12.

The topographical appearance was observed in repeated scans of the same area. Small streaks were found parallel to the probe movement on the material separating the islands. This together with the phase image pictures, indicate that the separating material is softer than the islands.

Table IV. Average Structure Size and Surface Roughness

Sample ID	Catalyst Concentration (W%)	Relative Humidity (%)	Molecular Weight (kD)	Average Structure Size (μm)	Surface Roughness (nm)
1	0.05	0	4.2	0.0	7
2	5	0	4.2	6.5	97
3	0.05	100	4.2	0.0	8
4	5	100	4.2	4.5	55
5	0.05	0	49	0.5	4
6	5	0	49	1.2	13
7	0.05	100	49	0.0	7
8	5	100	49	1.3	11
11	0.5	0	26	2.0	87
12	0.5	100	26	2.2	106
Centerpoints	0.5	53	26	1.4	24

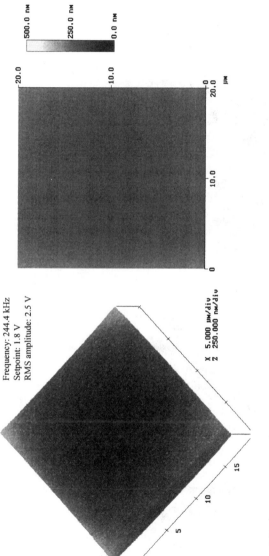

Frequency: 244.4 kHz
Setpoint: 1.8 V
RMS amplitude: 2.5 V

Figure 4. AFM image of sample ID 3

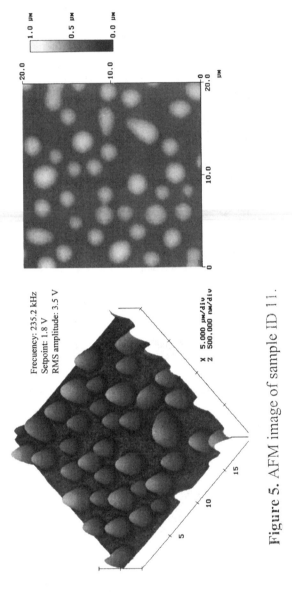

Frecuency: 235.2 kHz
Setpoint: 1.8 V
RMS amplitude: 3.5 V

Figure 5. AFM image of sample ID 11.

Fitting a model to the structure data revealed that a low molecular weight and high catalyst concentrations resulted in the largest structures, Figure 6. The formation of structures is dependent on the kinetics of the reaction and thermodynamics of the phase separation process. The structures formed are presumably composed of self-condensed FTEOS which have formed in the bulk and migrated to the surface or formed at the surface of surface-migrated uncross-linked FTEOS. The larger diameter of the structures when the catalyst concentration is increased corresponds with the results obtained with TEOS by Novak et. al. [20]. They achieved a ramified, elongated structure at low catalyst amounts and a larger densed structure when the catalyst concentration was increased. When the molecular weight of the polymer was raised the diameters of the structures decreased, which agrees with the results obtained for bulk structures by Black et al. [19]. The model explained about 85% of the variation (R^2). The predictability (Q^2) was 76%. ANOVA calculation showed no lack of fit and the normal probability plot of the residuals showed no outliers.

The phase-separated-induced surface structures observed in this study are comparable with the structures achieved by Karim et al. [8], which were observed in a copolymer blend of dPEP/PEP. The ondulations observed in their study originate from spinodal decomposition in the bulk. The phenomenon is general according to Karim et al. and may thus be the mechanism for the phase separated surface structures observed in our system.

Study of the surface morphology of sample ID 8 (Figure 7) revealed that a micelle-like structure has formed on the surface. This may be the fluorinated side-chain of FTEOS that has self-assembled. Self-assembly at the surface has previously been reported to occur in polymer systems with fluorinated side chains [21-22]. The migration of the low surface energy side chain, consisting of $-CF_2$ and $-CF_3$ groups, in the cross-linker now followed by a self-orientation at the surface may be responsible for the micelle-like structure formed. Further investigation with surface friction microscopy (SFM) could give additional useful information. Phase segregation and self-orientation could be tested and maybe verified with SFM.

An organization at the surface would give a more stable material concerning the surface energy in water environments. The surface energy of silicone coatings increases upon immersion in water due to molecular rearrangements of the silicone backbone, displaying the higher energy oxygen atoms at the surface [23]. The increased surface energy of silicone elastomers induced by water exposure may enable several interactions with the bioadhesive of marine macrofoulers and a stronger attachment of the organisms. A surface with negligible surface reconstruction, e.g. stable surface energy, when immersed in water would increase the practical applications of low surface energy polymers.

Conclusions

We have demonstrated that the surface chemistry and surface morphology of a sol-gel system composed of poly(dimethylsiloxane) (PDMS) and (tridecafluoro-1,1,2,2-

108

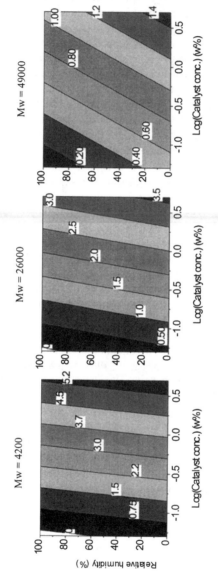

Figure 6. Contour plots based on the statistical model showing the average size (μm) of the structures at the surface.

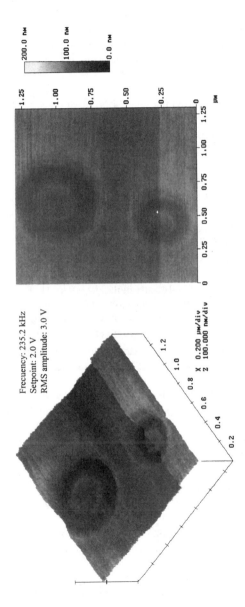

Frecuency: 235.2 kHz
Setpoint: 2.0 V
RMS amplitude: 3.0 V

Figure 7. AFM image of selected structures on sample ID 8

tetrahydrooctyl) triethoxysilane (FTEOS) cross-linker can be controlled by varying the catalyst concentration, molecular weight of PDMS and humidity during curing. Increasing the catalyst concentration and humidity both increased the amount of fluorine at the surface. By varying the catalyst concentration and molecular weight of PDMS the morphology of the micro-structures at the surface were controlled. The structures achieved in this investigation can be produced within the micro-size interval, which might be of critical importance for directing the biological response.

Acknowledgments

This work was supported by the Swedish Foudation for Strategic Research through the Marine Science and Technology (MASTEC) program.

Literature Cited

1. Chehroudi, B.; Gould, T. R.; Brunette, D. .M. *J. Biomedical Materials Research.* **1988**, *22*, 459.
2. von Recum, A. F.; van Kooten, T. G. *Journal of Biomaterials Science - Polymer Edition* **1995**, *7*, 181.
3. Britland, S.; Clark, P.; Connolly, P.; Moores, G. *Experimental Cell Research* **1992**, *198*, 124.
4. Healy, K. E.; Thomas, C. H.; Rezania, A.; Kim, J. E.; McKeown, P. J.; Lom, B.; Hockberger, P. E. *Biomaterials* **1996**, *17*, 195.
5. Muthukumar, M.; Ober, C. K.; Thomas, E. L. *Science* **1997**, *277*, 1225.
6. Straub, W.; Bruder, F.; Brenn, R.; Krausch, G.; Bielefeldt, H.; Kirsch, A.; Marti, O.; Mlynek, J.; Marko, J. F. *Europhysics Letters* **1995**, *29*, 353.
7. Walheim, S.; Boltau, M.; Mlynek, J.; Krausch, G.; Steiner, U. *Macromolecules* **1997**, *30*, 4995.
8. Karim, A.; Slawecki, T. M.; Kumar, S. K.; Douglas, J. F.; Satija, S. K.; Han, C. C.; Russell, T. P.; Liu, Y.; Overney, R.; Sokolov, O.; Rafailovich, M. H. *Macromolecules* **1998**, *31*, 857.
9. Jenekhe, S. A.; Chen, X. L. *Science* **1998**, *279*, 1903.
10. Ghadiri, M. R. *Advanced Materials* **1995**, *7*, 675.
11. Van der Grinten, M. G. D.; Clough, A. S.; Shearmur, T. E.; Bongiovanni, R.; Priola, A. *Journal of Colloid and Interface Science* **1996**, *182*, 511-515.
12. Chapman, T. M.; Benrashid, R.; Marra, K. G.; Keener, J. P. *Macromolecules* **1995**, *28*, 331.
13. Swain, G. W.; Schultz, M. P. *Biofouling* **1996**, *10*, 187.
14. Watermann, B.; Berger, H. D.; Sonnichsen, H.; Willemsen, P. *Biofouling* **1997**, *11*, 101.
15. Johnston, E.; Gatenholm, P.; Bullock, S.; Wynne, K. J. *Abstracts of Papers of the American Chemical Society* **1998**, *39, 510-511.*
16. Johnston, E.; Bullock, S.; Uilk, J.; Gatenholm, P. and Wynne, K. J. *Macromolecules* **1999**, *32*, 8173 -8182.

17. ASTM E 104-85, American Society for Testing and Materials. *Standard Practice for Maintaining Constant Relative Humidity by Means of Aqueous Solutions*, **1991**.

18. Iwamoto, T.; Morita, K.; Kackenzie, J. D. *Journal of non-crystalline solids* **1993**, *159*, 65.

19. Black, E. P.; Ulibarri, T. A.; Beaucage, G.; Schaefer, D. W.; Assink, R. A.; Bergstrom, D. F .; Giwa-Agbomeirele, P. A.; Burns, G. T. In *Hybrid Organic-Inorganic Composites;* Mark, J. E; Lee C. Y-C.; Bianconi, P. A., Eds; ACS Symposium Series 585; American Chemical Society: Washington D.C, 1995, 237

20. Novak, B. M.; Ellsworth, M. W.; Verrier, C. In *Hybrid Organic-Inorganic Composites;* Mark, J. E; Lee C. Y-C.; Bianconi, P. A., Eds; ACS Symposium Series 585; American Chemical Society: Washington D.C, 1995, 237

21. Wang, J.; Mao, G.; Ober, C. K. and Kramer, E.J. *Macromolecules* **1997**, *30*, 1906.

22. Y. Katano, H. T. and T. Nakajima *Macromolecules* **1994**, *27*, 2342.

23. Owen, M. J. *Ind. Eng. Chem. Prod. Res. Dev.* **1980**, *19*, 97-103.

Chapter 9

Wettability of Fluorosilicone Surfaces

Susan V. Perz, Christopher S. McMillan, and Michael J. Owen

Dow Corning Corporation, 2200 West Salzburg Road,
Midland, MI 48686–0994

Radio frequency plasma treatment of polydimethylsiloxane (PDMS) is a useful way of increasing wettability to improve adhesion. Its main defect is the subsequent rapid recovery of hydrophobicity. The fluorosilicone polymethyltrifluoropropylsiloxane (PMTFPS) is another low surface energy silicone where improved wettability is often desired. We have directly compared the behavior of PMTFPS and PDMS using air, oxygen, helium and argon plasma gases. The effect of the plasma has been investigated by water contact angle changes and by x-ray photoelectron spectroscopy (XPS) surface characterization. Both an unfilled PMTFPS gum and a filled elastomer were examined. PMTFPS is affected in much the same way as PDMS, an oxidized silica-like surface region is produced. This is shown by the shift in the high resolution Si 2p spectra to higher binding energy which is most marked in the case of helium treatment. Significant improvements in water wettability occur with helium treatment having the greatest effect but the hydrophobic recovery is mostly complete within 24 hours in all cases. Multiple treatments followed by water storage are effective in maintaining wettability for at least a week.

Adhesion to low surface energy polymers based on fluorocarbons, hydrocarbons and silicones is often difficult to achieve satisfactorily and silicones are no exception. These polymers derive their low surface energy from aliphatic alkyl or fluoroalkyl groups; the two most familiar silicone examples of these two classes of polymer are polydimethylsiloxane (PDMS) and polymethyltrifluoropropylsiloxane (PMTFPS). The chosen polymer is usually selected for a particular desired bulk attribute such as chemical inertness, thermal stability, flexibility at low temperature, gas permeability, and so on. Composite multilayer films containing such polymers are often required and good adhesion between the layers is essential. The low surface energy makes it impossible for most adhesives to adequately wet these polymers, moreover, the intermolecular forces involved at the adhesive/polymer interface are often dominated by the weak London forces inherent to aliphatic hydrocarbon and fluorocarbon groups. These properties provide commercial opportunities as release and antifouling coatings, for example, but there are many situations where enhanced adhesion to silicones is very desirable. In addition to seeking enhanced wettability as the essential first step in achieving good bondability of polymers to adhesives and other substrates, there are also other applications, particularly in the biomedical material area, where increased wettability is the primary goal of the treatment. Examples include tissue culture surfaces with improved cell attachment and growth characteristics, and contact lens material surfaces with enhanced wettability by tears and altered protein and lipid deposition.

Chemical pretreatments have been developed to enhance adhesion to fluorocarbon polymers (*1*) such as polytetrafluoroethylene (PTFE). The main commercial approaches use sodium in liquid ammonia and sodium naphthalenide in tetrahydrofuran. Potassium hydroxide has also been shown to be effective for other fluoropolymers such as poly(vinyl fluoride) and poly(vinylidene fluoride). These treatments have some negative environmental consequences from the use of volatile solvents and reactive intermediates, as well as by-product waste-stream problems. There is also the specific disadvantage for fluorosilicones that certain of these chemical treatments can adversely affect the siloxane backbone. More environmentally acceptable surface oxidizing treatments such as plasma, corona and flame treatment have been shown to be effective for a variety of fluoropolymers and would seem to be worth applying to fluorosilicones.

Plasma modification of polymers, first introduced commercially almost thirty years ago, is still growing in importance and is currently receiving considerable attention both in industry and academe. As only a few molecular layers in the surface region are usually changed by the plasma it provides a useful way of controlling surface properties and not affecting the bulk properties of the polymer without the drawbacks of chemical pretreatments. Such treatments are not devoid of drawbacks of their own, for example, excessive treatment can lead to detrimental surface damage. A particular problem with plasma or corona treatment of common siloxane polymers based on PDMS is the rapid recovery of hydrophobicity on cessation of the treatment (*2*).

Because of its central importance in the silicone industry, the effect of plasma treatment on PDMS has been extensively investigated (*2-7*). A wide variety of plasma gases has been used including air, water vapor, argon, helium, nitrogen, oxygen, ammonia and carbon dioxide. Also varying times, powers, pressures, gas flow rates and other conditions as well as differences in excitation frequency (radio frequency (RF) or microwave), have been employed in these studies. Expectedly, there are significant differences in behavior reported, but in general for the RF plasma treatments, a thin wettable silica-like layer is produced. This layer is brittle and microcracking is evident at more severe levels of plasma treatment. Conditions can be found to achieve adequate wettability without microcracking but the more difficult drawback is the progressive loss of wettability after cessation of the treatment. This hydrophobic recovery results either from reorientation of untreated methyl groups back into the surface replacing plasma-produced surface hydrophilic groups (reorientation or "overturn" mechanism) or migration of treated polymer chains into the bulk or untreated polymer chains from the bulk to the surface (diffusion mechanism). Hydrophobic recovery of PDMS is often marked within twenty four hours and virtually complete within one week.

The plasma treatment of PMTFPS has not been previously investigated despite its significance as another hydrophobic, low-surface-energy silicone polymer. More highly fluorinated fluorosilicones exist but PMTFPS is by far the most widely available. It combines excellent solvent resistance with adequate thermal stability for the minimum degree of fluorination. Essentially, it contains the least amount of fluorine (least added cost) consistent with marked solvent resistance improvement over PDMS. PMTFPS elastomers, sealants and coatings are particularly well suited to the many applications involving exposure to fuels, lubricants, hydraulic fluids, and solvents. Compared to other fuel-resistant polymers, PMTFPS materials offer the widest operating service temperature range and hardness range of any material. These attributes, coupled with the general ease of fabrication, make PMTFPS an increasingly frequent candidate for incorporation in composite assemblies where good adhesion is essential. This plasma modification study of PMTFPS was undertaken to determine if the wettability of PMTFPS could be improved in this way, how it compared to the widely investigated PDMS case, whether drawbacks such as hydrophobic recovery were also experienced, and also to evaluate some ways of retaining the hydrophilic surface that had been suggested for PDMS such as repeated plasma treatments (3) and storage in water (4).

Surface Properties of PMTFPS

Surface energy is the fundamental surface property of polymers. It can be considered either a surface tension (force per unit length, mN/m) or a surface free energy (free energy per unit area, mJ/m^2). These quantities are identical for liquids but not for solids. Table I shows a variety of surface energy and water contact angle values of PMTFPS and comparable hydrocarbon, fluorocarbon and silicone polymers. Water contact angles are included as a direct index of hydrophobicity. The data are taken from Physical Properties of Polymers Handbook (8) and a recent fluorosilicone review (9). The liquid surface tensions, extrapolated to infinite MW, and the water contact angles are directly measured, whereas the solid surface energies are inferred from contact angle measurements using semi-empirical equations with values dependent on the choices made. Three approaches are given in Table I. The polar and dispersion force components come from the Owens and Wendt geometric mean approach using water and methylene iodide. It is this latter liquid that behaves very differently on PDMS and PMTFPS and accounts for the differences shown in Table I. The other dispersion force component column comes from the Girifalco-Good-Fowkes approach using n-hexadecane. Also included are Zisman critical surface tensions of wetting. These different choices of semi-empirical equations and contact angle liquids for approximating surface energy account for the differences seen in the table. These data are provided to establish that PMTFPS is a polymer of comparable low surface energy and hydrophobicity to such more familiar polymers as PDMS and PTFE.

Table I: Surface Properties of Low Surface Energy Polymers

Polymer	γ_{LV} [a] (mN/m)	γ^d [b] (mN/m)	γ^p [c] (mN/m)	γ_s [d] (mN/m)	$\gamma^d(H)$ [e] (mN/m)	γ_C [f] (mN/m)	$\theta\ H_2O$ [g] (deg)
PMTFPS	24.4	10.8	2.8	13.6	18.3	21.4	104
PDMS	21.3	21.7	1.1	22.8	22.6	24	101
PTFE	25.6	18.6	0.5	19.1	19.8	18.3	108

(a) Liquid surface tension extrapolated to infinite MW
(b) Dispersion force component by Owens and Wendt geometric mean approach
(c) Polar component by Owens and Wendt approach
(d) Solid surface tension, sum of previous two figures
(e) Dispersion force component from Girifalco/Good/Fowkes approach
(f) Zisman critical surface tension of wetting
(g) Water contact angle
[See reference 8 for further details and explanations]

Experimental

Quasi-static, advancing contact angles were measured using doubly-distilled, deionized water with a Video Contact Angle (VCA) 2000 System manufactured by Advanced Surface Technology, Inc. Values presented in Figures 1 and 2 and Table II are the average of six readings on both sides of three drops. The precision of measurement of an individual value is better than 0.1 deg but considering the nature and reproducibility of elastomer surfaces the accuracy is *circa* +/- 5 deg. X-ray photoelectron spectroscopy (XPS) analyses were obtained using a Kratos AXIS 165 instrument with monochromatic Al x-ray source at 240 W power, charge compensation employed and analysis spot size *circa* 400 x 800 microns. The instrument precision for the atomic compositions presented in Figures 4 and 5 and Table III is 0.1% and the accuracy of the measurements is *circa* 0.5%. Plasma treatments were performed in a Branson/IPC S4000 Series plasma system which generates a low pressure, radio frequency (13.56 MHz), cold plasma. Controllable parameters include treatment time, gas type, gas flow-rate, RF power, pressure and position of sample in chamber. In this study the first two were varied and the rest kept as constant as possible. Gas flow rate was set at 50 (units unknown) on the instrument flow meter and RF power was 150 Watts. Some variation in pressure was required to maintain the discharge with different gases at this constant power. We studied air, oxygen, helium and argon; pressures varied from 1 to 3 torr and are given for each gas in Table II. Because of these many variables we included a PDMS control in our

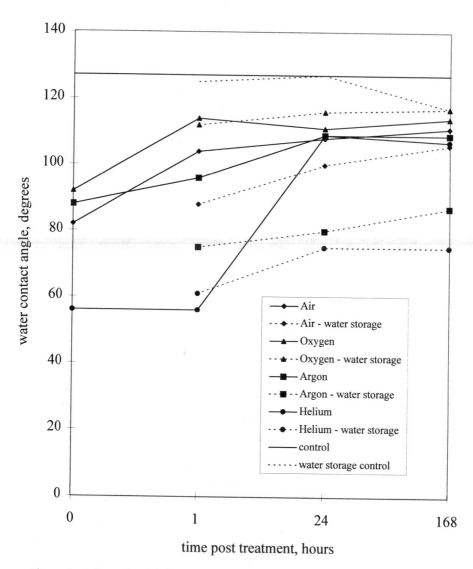

Figure 1. Effect of Multiple Plasma Treatments and Water Storage on PMTFPS Gum

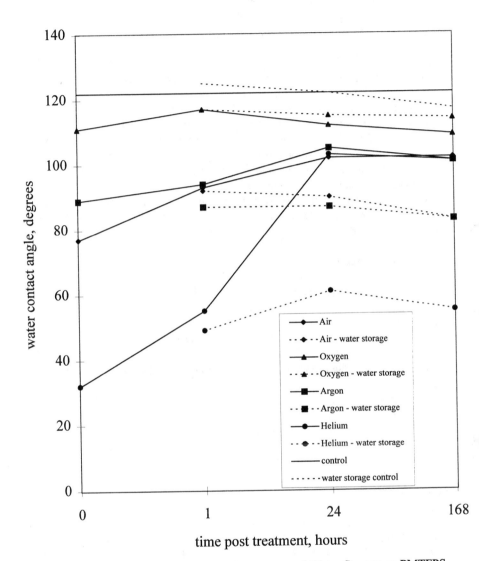

Figure 2. Effect of Multiple Plasma Treatments and Water Storage on PMTFPS Elastomer

present studies for comparison and have made no attempt to relate these data directly to other PDMS data in the literature. Samples for XPS evaluation of plasma treatment were treated for 5 minutes and analysis begun approximately 45 minutes later when the considerable outgassing of samples that occurred had subsided to tolerable levels (*circa* 2×10^{-8} torr pressure in the analysis chamber).

Two PMTFPS materials were examined, an unfilled polymer gum and an elastomer formulated from that gum by the incorporation of fumed silica filler. The gum is a DC internal intermediate known as SGM 900 and the elastomer is a commercial product designated LS 2840. The PDMS elastomer comparison, also silica filled, is known as X7-5203. Some preliminary experiments with solvent wiping showed no difference in behavior so elastomer samples were used as received. For post-treatment aging the samples were covered but not otherwise isolated from the laboratory environment. Previous studies of these PMTFPS materials (*10*) have shown that uncontaminated surfaces of the unfilled gum can usually be obtained but that the filled elastomers are usually slightly contaminated with PDMS. This is likely to be a surface contaminant and probably results from compounding the elastomer in an industrial environment where PDMS is the major polymer processed.

Results and Discussion

Quasi-equilibrium advancing contact angles of water on the plasma treated materials are given in Table II. 10 second treatments were also studied but had no significant effect on the contact angle. The behavior of PMTFPS and PDMS is broadly the same. Aberrations in the data, such as 60 second exposure occasionally giving a higher contact angle than 30 second exposure, are probably due to sample and treatment fluctuations. For instance contact angles were taken on different parts of the samples which may be of different roughness. These relatively short times were used to ensure that the surface was not physically damaged by the plasma leading to much greater uncertainties than our estimated +/- 5 deg accuracy of contact angle measurement. Exposure to these plasma conditions for several minutes produced similar microcracking to that experienced with PDMS (*2*). Clearly, both polymers exhibit marked hydrophobic recovery with almost all the recovery occurring within twenty-four hours. Samples were re-examined at 7 days and generally showed no significant further change.

Comparisons between the two polymers should be made with the data on filled systems. There is an indication that under the same conditions the PMTFPS becomes more wettable than PDMS. This is in line with their high temperature, thermo-oxidative stability behavior. Interestingly, the unfilled PMTFPS gum is made more wettable than the filled elastomer implying involvement of the silica filler in the plasma oxidation.

Figures 1 and 2 show the effect of multiple treatments and water storage for the PMTFPS gum and elastomer, respectively. The samples were treated for 60 seconds three days in a row and the water contact angle monitored for samples stored in air and in water. As with the single 60 second treatments, helium stands out as the most effective plasma gas for enhancing wettability. Multiple treatments followed by air storage are not very effective in preserving wettability, there are indications at one hour aging that multiple helium treatments help but by 24 hrs all materials with all plasmas show marked hydrophobic recovery to water contact angles of 100 deg or greater. Water storage is more effective, particularly for the fluorosilicone elastomer. At best a *circa* 50 deg water contact angle can be maintained for one week with multiple helium treated, water stored LS 2840.

Table II: Effect of Plasma Treatment Time on Wettability

Polymer	Plasma Gas	Time (s)	Water Contact Angle (deg)	
			Initial	After 24 hr
PMTFPS gum	Air	30	57	107
		60	44	114
	Oxygen	30	96	112
		60	78	110
	Helium	30	58	100
		60	41	101
	Argon	30	46	111
		60	80	114
PMTFPS elastomer	Air	30	86	101
		60	90	101
	Oxygen	30	101	105
		60	76	105
	Helium	30	89	102
		60	50	98
	Argon	30	70	104
		60	60	106
PDMS elastomer	Air	30	105	103
		60	98	103
	Oxygen	30	110	119
		60	61	98
	Helium	30	77	96
		60	81	99
	Argon	30	82	99
		60	90	95

Table III summarizes the XPS surface composition of the untreated PMTFPS materials. The gum is seen to be very close in composition to theoretical expectation whereas the filled elastomer seems considerably contaminated, its surface region approximating much more closely to a 1:1 PMTFPS/PDMS mixture than to the PMTFPS gum. The high resolution C 1s spectrum is interesting, see Figure 3 for PMTFPS gum, particularly for silicone surface scientists used to the single C 1s peak of PDMS. Three peaks are seen with binding energies (BE, referenced to 284.6 eV for

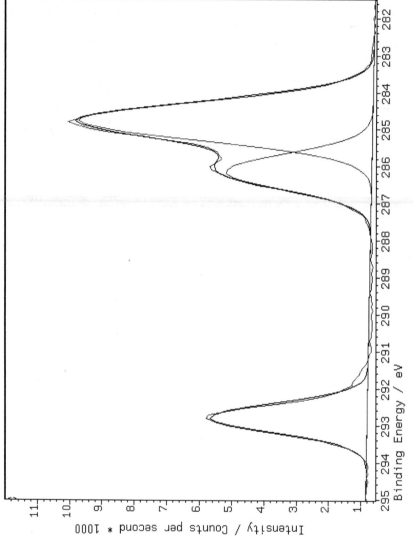

Figure 3. C 1s Spectrum for PMTFPS Gum

the lowest BE peak) at 292.6 eV, 286.0 eV and 284.6 eV. The 292.6 eV peak is the CF_3 carbon, the 286.0 eV peak is the CH_2 carbon adjacent to the CF_3 and the 284.6 eV peak is the CH_2 adjacent to silicon and the CH_3 peak. As Figure 3 shows these are present in approximately the expected 1:1:2 ratio (1 + 2 : 3 = 0.85, not the theoretical 1.0 so a little PDMS contamination may be present even on the gum surface). This secondary binding energy shift of the carbon atoms adjacent to those substituted is well established in XPS studies of fluoropolymers. For example, comparable binding energies for poly(1,1-difluoroethylene), $\{(CF_2CH_2)_n\}$, are 290.8 eV and 286.3 eV (referenced to ethylene at 285.0 eV) (11).

Table III: XPS Surface Composition of Untreated Samples

Sample	Atomic Composition (%)			
	F	C	O	Si
PMTFPS Gum	33.9	43.2	13.5	9.5
PMTFPS Gum [10]	38.1	40.7	11.1	10.1
PMTFPS Theory	33.3	44.4	11.1	11.1
PMTFPS Elastomer	18.4	44.3	20.5	16.8
1:1 PDMS/PMTFPS Theory	16.7	47.2	18.1	18.1

The effect of plasma exposure is shown diagrammatically in Figure 4 for the PMTFPS gum and in Figure 5 for the PMTFPS elastomer. For the gum fluorine and carbon contents drop to about a third of their original levels in helium plasma while oxygen and silicon levels roughly triple. Effects are less marked with the other plasma gases but change in the same direction. Other detailed differences are noticeable in the XPS study. For example, the different treatment gases produced samples requiring rather different pumpdown times, suggesting different abilities to produce low MW fragments. The filled elastomer surfaces for each plasma gas closely resemble the corresponding surfaces of the plasma treated gum despite the considerable differences in the untreated surfaces composition, so the presumed PDMS surface contamination has no effect on the plasma oxidation. For both the filled and unfilled materials, a shift in the high resolution Si 2p spectra, shown in Figure 6 for the gum and in Figure 7 for the elastomer, to higher binding energy, most marked in the case of helium treatment, is strongly indicative of the formation of a silica-like layer as is observed with PDMS RF plasma treatment.

Figure 8 is the C 1s high resolution spectrum for the PMTFPS gum treated in helium, other gases and the elastomer gave broadly similar results. The reduction in the BE peaks ratio (1 + 2 : 3) from 0.85 untreated to 0.64 treated (compare Figure 8 with Figure 3) implies the fluoroalkyl $CF_3CH_2CH_2$- group is more affected by the plasma treatment than the methyl groups, as was suggested by the contact angle data. These data also imply a possible role for the silica filler in the plasma treatment. Silica filler is present in the surface regions as can be seen from Figures 6 and 7. Figure 6 contains the high resolution Si 2p spectrum for the unfilled gum and Figure 7 the same for the filled elastomer. The considerable broadening on the high binding energy

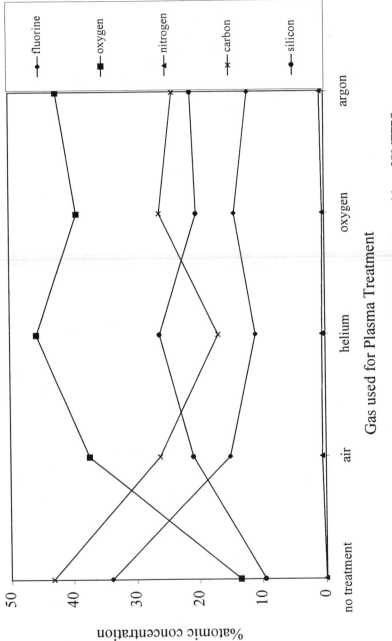

Figure 4. Effect of Plasma Treatment on XPS Surface Composition of PMTFPS Gum

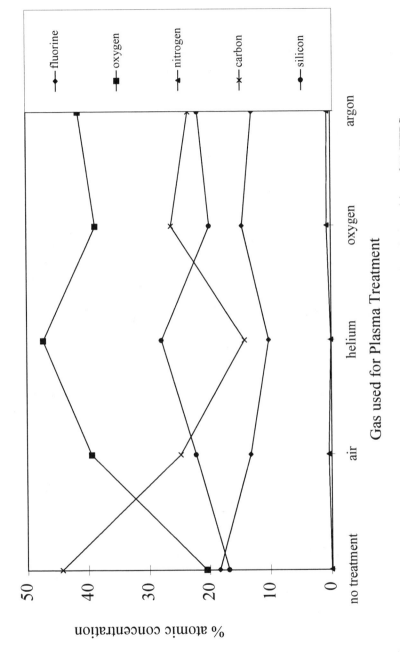

Figure 5. Effect of Plasma Treatment on XPS Surface Composition of PMTFPS Elastomer

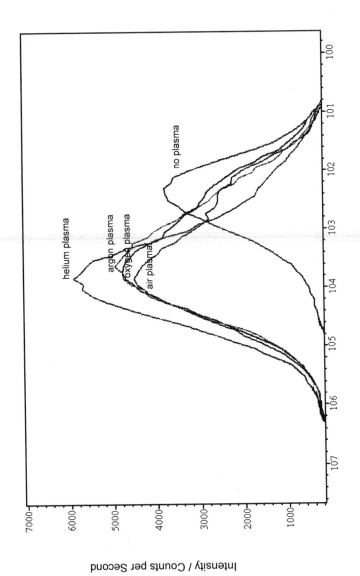

Figure 6. Si 2p Spectra of Plasma Treated and Untreated PMTFPS Gum

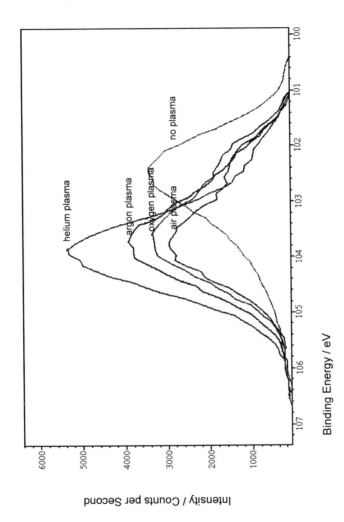

Figure 7. Si 2p Spectra of Plasma Treated and Untreated PMTFPS Elastomer

126

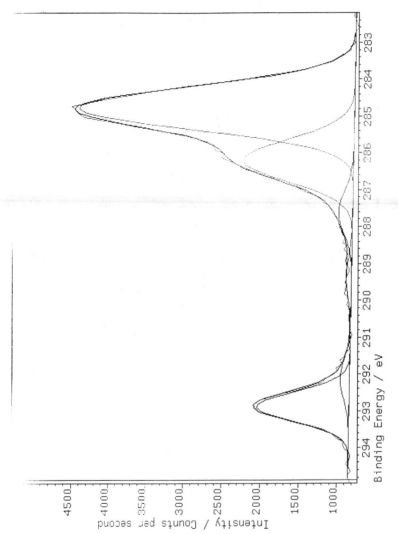

Figure 8. C 1s Spectrum for Helium Plasma Treated PMTFPS Gum

side of the latter spectrum is indicative of the presence of filler in the surface region and offers an additional explanation for the divergence of surface atomic composition data for the elastomer given in Table III. This behavior is different to PDMS elastomers where indications of filler are rarely seen with surface sensitive techniques such as XPS.

Summary

PMTFPS is affected by RF plasma treatment in much the same way as PDMS. Significant improvements in water wettability are possible but the treatments are temporary and hydrophobic recovery is mostly complete within 24 hours. Detailed differences are evident with various plasma gases; helium generally having the most significant effect on PMTFPS wettability. There are also differences in hydrophobic recovery behavior of filled and unfilled PMTFPS materials. Multiple treatments followed by water storage are effective in maintaining wettability for at least a week.

XPS studies of the untreated PMTFPS materials show good agreement with expected composition for the unfilled gum but considerable differences for the filled elastomer. This seems to be due to a combination of some PDMS surface contamination and the presence of silica filler in the surface region. These factors do not significantly affect the outcome of plasma treatment under the conditions we have chosen. For a given plasma gas, the filled and unfilled PMTFPS samples are affected similarly although there are significant differences between the various gases. XPS studies also show that helium has the greatest effect of the plasma gases used in this study. For both the filled and unfilled materials, a shift to higher binding energy of the Si 2p spectrum strongly suggests the formation of a silica-like surface region. Similar surface oxidation has been reported for the case of PDMS.

These initial feasibility studies suggest that the higher glass transition temperature of PMTFPS than PDMS (193 K compared to 146 K) does not result in significantly slower hydrophobic recovery. It may be that more low MW diffusible cyclic species in the PMTFPS compensate but this is the subject of future studies. Certainly nothing in this present study enables us to distinguish between the competing surface reorientation and diffusion mechanisms that have been proposed as the cause of the hydrophobic recovery effect.

Literature Cited

1. Brewis, D. M.; Mathieson, I. In *Modern Fluoropolymers*, Schiers, J., Ed.; Wiley: New York, 1997; p. 165.
2. Fritz, J. L.; Owen, M. J. *J. Adhesion*, **1995**, *54*, 33.
3. Morra, M.; Occhiello, E.; Marola, R.; Garbassi, F.; Humphrey, P.; Johnson, D. *J. Colloid Interface Sci.*, **1990**, *137*, 11.
4. Everaert, E. P.; Van der Mei, H. C.; De Vries, J.; Busscher, H. J. *J. Adhesion Sci. Technol.*, **1995**, *9*, 1263.
5. Triolo, P. M.; Andrade, J. D. *J. Biomed. Mater. Res.*, **1983**, *17*, 129.

6. Hettlich, H-J.; Otterbach, F.; Mittermayer, Ch.; Kaufmann, R.; Klee, D. *Biomaterials*, **1991**, *12*, 521.
7. Owen, M. J.; Smith, P. J. *J. Adhesion Sci. Technol.*, **1994**, *8*, 1063.
8. Owen, M. J. In *Physical Properties of Polymers Handbook*, Mark, J. E., Ed.; AIP Press: Woodbury, NY, 1996; p. 669.
9. Maxson, M. T.; Norris, A. W.; Owen, M. J. In *Modern Fluoropolymers*, Scheirs, J., Ed.; Wiley: New York, 1997; p. 359.
10. Owen, M. J. *J. Appl. Polym. Sci.*, **1988,** *35*, 895.
11. Clark, D. T.; Feast, W. J.; Kilcast, D; Musgrave, W. K. R. *J. Polym. Sci.*, **1973** *11*, 389.

Chapter 10

Design of Fluorinated Acrylic-Based Polymers as Water Repellent, Intrinsically Photostable Coating Materials for Stone

V. Castelvetro[1], M. Aglietto[1], F. Ciardelli[1], O. Chiantore[2], and M. Lazzari[2]

[1]Dipartimento di Chimica e Chimica Industriale, Università di Pisa, Via Risorgimento 35, 56126 Pisa, Italy
[2]Dipartimento di Chimica Inorganica, Chimica Fisica e Chimica dei Materiali, Università di Torino, Via P. Giuria 7, 10125 Torino, Italy

Several copolymers of fluorinated acrylates and methacrylates with unfluorinated acrylates, methacrylates and vinyl ethers have been synthesized, and thin films of these materials have been evaluated in terms of chemical stability under accelerated photoaging conditions. Either commercially available or specifically synthesized fluorinated acrylic comonomers, such as methyl α-trifluoromethyl. acrylate (MTFMA) and ethyl 3,3-difluoro-2-methylpropenoate (MFPE), were employed. If suitably selected, these fluoroacrylates can allow both control on the macromolecular structure and tailoring of material properties that are relevant for coating applications, such as film-forming behavior, adhesion, water repellency and intrinsic photostability. The photodegradation behavior of partially fluorinated copolymers and of their unfluorinated analogues was compared, and specific structural features responsible for crosslinking and/or bond scission reactions were analysed.

Fluorinated polymers have gained growing popularity in the last 10 years, particularly as the consequence of an increasing industry-pulled demand for high performance, durable materials with unusual interfacial properties. Partially fluorinated coating materials and, among them, those based on fluorinated acrylic-based copolymers, are more and more frequently employed as components for paint and varnish formulations, in the textile and construction industry, i.e., in application fields considered to be characterized by mature technology.

Specifically designed introduction of fluorinated groups in the structure of either fully acrylic or acrylic-vinyl ether copolymers can be easily achieved through the use of commercially available or newly synthesized fluorinated acrylic comonomers (*1*). If suitably selected, these fluoroacrylates can allow both control on the macromolecular structure and fine tuning of material properties, that can therefore be tailored for specialized applications. In the production of coating materials, for example, the requirements in terms of film-forming behavior, mechanical and optical properties, adhesion, water repellency and intrinsic photostability can be very different, depending on the specific application field and technique, on the nature of the coated substrate and on the conditions of weather exposure.

In particular, fluorine content and distribution along the macromolecular structure can be easily modified by selecting suitable comonomers and polymerization conditions (*2-6*). An appropriate selection of the partially fluorinated acrylate or methacrylate in radical initiated copolymerizations with unfluorinated acrylates, methacrylates and/or vinyl ethers, can allow the synthesis of either random or alternating copolymers; block copolymers could be prepared, in principle, from the same monomers by using the recently developed living radical polymerization techniques. These differences in comonomer distribution along the macromolecular chain can have a dramatic influence on the chemical stability and physical-chemical properties of the polymeric material.

The versatility of the synthetic approach is well suited for the development of new products for high value-added sectors, as in the case of protective coating materials for the safeguard of cultural heritage. In recent years, it has been recognized that most, if not all, the products employed in the past for the consolidation and protection of stone objects of art and monuments were "short lasting" and even detrimental to the preservation of the stone surface (*7,8*), often contributing, in the long term, to the progression of stone deterioration due to weathering. In Italy, where restoration and preservation of monuments and façades of historic buildings is usually preferred to the replacement of the deteriorated parts, the Committee for Stone Material Normalization (NORMAL) has outlined the fundamental requirements for a protective coating material (*9*). These are impermeability to liquid water, good permeability to water vapour, chemical and photochemical stability, inertness towards the stone substrate, reversibility of the treatment, good optical properties (i.e., negligible short and long term modification of the appearance of the stone, in terms of colour and gloss). Acrylic polymers and their mixtures with silicones are still widely employed in the consolidation and protection of deteriorated stone surfaces; however these products were not specifically developed for this purpose, and the photostability of acrylic polymers is actually far from sufficient, leading to loss of efficacy of the treatment and to undesired effects on the appearance of treated surface (yellowing, etc.). In addition, the use of reactive silicones, that undergo crosslinking and form chemical bonds with the stone substrate, does not allow subsequent removal of the product, if required in future restoration works.

In order to develop new materials that could be more successfully used as protective coatings and consolidants for valuable stone surfaces, better knowledge of the coating behavior, stability and penetration inside stones of highly variable chemical composition and porous structure are needed.

For this purpose, in the current investigation several copolymers of fluorinated acrylates and methacrylates with unfluorinated methacrylates and vinyl ethers have been synthesized and evaluated in terms of chemical and photochemical stability. Their behavior as protective coating materials was studied after application to various stone substrates that are commonly found in ancient buildings and objects of cultural significance, and compared to that of their non-fluorinated homologues (10). Here a set of acrylates and methacrylates with short and long fluorinated alkoxy chains, and/or with either fluorine or trifluoromethyl substituted vinyl group, have been employed in order to assess the influence of fluorine on the photodegradation pathway as determined upon accelerated photoaging experiments under strictly controlled conditions. Comparison with the behaviour of not fluorinated homologues (e.g. copolymers of ethyl methacrylate vs. those of 2,2,2-trifluoroethyl methacrylate) has also been used in this case in order to correlate the observed photodegradation behaviour with the chemical structure of the macromolecule.

Experimental

Materials. 2,2,2-trifluoroethyl methacrylate (TFEM), 1H,1H,2H,2H-perfluorodecyl methacrylate (XFDM) and acrylate (XFDA), were prepared from the corresponding alcohols and acryloyl- or methacryloyl chlorides. The commercial products butyl vinyl ether (BVE), 2-ethylhexyl vinyl ether (EHVE), ethyl methacrylate (EM), butyl methacrylate (BM), 1,1,1,3,3,3-hexafluoroisopropyl methacrylate (HFIM) and α-fluoro-1,1,1,3,3,3-hexafluoroisopropyl acrylate (HFIFA) were distilled before use. Lumiflon LF200 (Zeneca) was used as supplied. Methyl α-trifluoromethyl acrylate (MTFMA) and ethyl 3,3-difluoro-2-methylpropenoate (MFPE) were supplied by Prof. U. Matteoli and C. Botteghi, from the Dept. of Chemistry of the University of Venice (Italy). Copolymer compositions were determined by ^{1}H-NMR and their structure by FT-IR, SEC, ^{1}H, ^{13}C and ^{19}F NMR.

| TFEM | XFDM | HFIM | MTFMA | HFIFA | MFPE |

Techniques. DSC analyses were carried out with a Perkin Elmer DSC7 instrument equipped with a liquid nitrogen CCA7 temperature controller for subambient operations. NMR spectra were recorded from CDCl₃ solutions on a Varian Gemini 200 spectrometer at 200 Mhz. SEC analyses were carried out in THF using a Waters M45 pump, a Rheodyne 7110 injector and four PL-Gel columns (500 - 10^5 Å); a series of poly(methyl methacrylate) standards and Bruker Chromstar software were employed for Mw and Mn calculations. FT-IR spectra were recorded on a Perkin Elmer 1710 spectrophotometer.

Polymer Synthesis. All copolymers (Table I) were prepared by conventional AIBN initiated radical polymerization, either in bulk (MTFMA/EHVE copolymer) or in solution (dioxane or toluene), and purified twice by precipitation from the appropriate solvents.

Table I. Fluorinated copolymers and unfluorinated analogues

Polymer	Molar Composition	F (wt %)	Tg (°C)	$\overline{Mn}$ $(x10^{-4})$	$\overline{Mw}/\overline{Mn}$
TFEM/BM	60/40	22	68	2.55	3.3
TFEM/BVE	77/23	29	32	3.14	1.6
EM/BVE	71/29	/	n.d.	2.00	3.4
XFDM/BA	17/83	28	-34	3.58	1.4
XFDA/BM	18/82	28	-13	4.06	1.7
HFIM/BVE	53/47	35	21	2.25	2.1
MTFMA/EHVE	60/40 [a]	22	43	1.79	3.3
MFPE/BVE	51/49	15	21	2.17	4.1
HFIFA/BVE	50/50	39	35	17.4	2.8
Lumiflon LF200	n.d.	n.d.	38	1.40	3.5

[a] Bulk polymerization, equimolar feed composition

Accelerated Photoaging. Thin (20-50 μm) films of the polymeric materials were obtained by casting from 5 wt% CHCl₃ solutions on untreated glass slides, followed by vacuum drying at 65°C. Additional samples of thinner (10-20 μm) films were laid on quartz slides and silicon wafers for FT-IR and UV-Vis analyses. Irradiation of the polymers was carried out in a Suntest CPS Heraeus chamber equipped with a Xenon lamp and 295 nm cutoff filter, to simulate the solar spectrum. Constant temperature (44°C) was maintained in the chamber by means of controlled ventilation, and irradiation was kept at 765 W/m² for as long as 2200 h, depending on the photostability shown by the polymeric material. The progress of photoinduced chain scission and/or crosslinking reactions was followed by gravimetry and SEC analysis, while the structural changes due to photooxidative degradation were monitored by IR and UV-Vis spectroscopy.

Results and Discussion

After a preliminary screening, which was carried out in order to assess the behavior of the various fluorinated monomers in radical initiated copolymerization with either unfluorinated acrylates, methacrylates or vinyl ethers (6), a set of copolymers reported in Table I was selected for the studies of photodegradation and water permeability. The XFDM and XFDA copolymers were compared in order to study the influence of a photolabile (tertiary) hydrogen atom, when present in either a fluorinated side chain or an unfluorinated repeat unit. The couple TFEM/BVE vs. EM/BVE can give information on the photostabilizing influence of side chain fluorinated methacrylic units in copolymers with vinyl ethers. TFEM/BVE and TFEM/BM can give indications on the relative photostability of vinyl ether vs. methacrylate copolymers with TFEM. The HFIM, MTFMA, HFIFA, and MFPE copolymers with vinyl ethers allow us to study how the degradation pathways can be influenced by fluorinated substituents (F or CF_3) at different positions of the side chain and/or of the backbone. Finally, Lumiflon LF200, a commercial copolymer containing $-CF_2-CFCl-$ units alternating with alkyl or hydroxyalkyl vinyl ethers, is used as a reference material for copolymers of vinyl ethers.

Previous results on homo- and copolymers of TFEM and XFDM with methacrylates having either branched or linear side chains had shown that the presence of a fraction of partially fluorinated side chains can slow down the photodegradation process, which is however still dominated by the reactivity of the unfluorinated units (10). In particular branched or linear but long unfluorinated side chains contribute to the progressive gelification of the material due to photoinduced crosslinking reactions that take place in the presence of atmospheric O_2.

Production of crosslinked, insoluble polymer, being an unwanted property for the protective coatings used in safeguarding of cultural heritage, may be used as one indicator for performance assessment. The gel content at different irradiation times was determined for most of the investigated polymers, and the results are reported in Figure 1. In all the samples containing vinyl ether units crosslinking is much lower, in comparison with the other copolymers. Moreover, it is further verified that crosslinking rate is large when acrylate or methacrylate units have long unfluorinated side chains.

The two XFDM and XFDA copolymers show a comparable extent of insoluble polymer fraction induced by the photooxidative treatment. The tendency of these structures to undergo a degradation pathway where gel formation is a main output may be revealed, even at relatively short degradation times, by looking at the changes in the size exclusion chromatography curves of the polymers. In fact, it is seen in Figure 2 that at the beginning of the degradation the molecular weight distributions

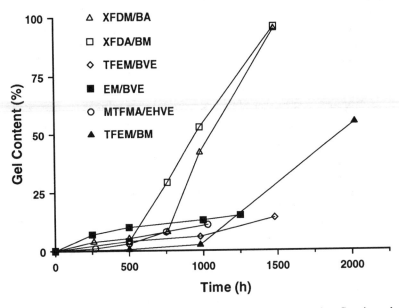

Figure 1. Gel content of selected fluorinated copolymers and unfluorinated analogues.

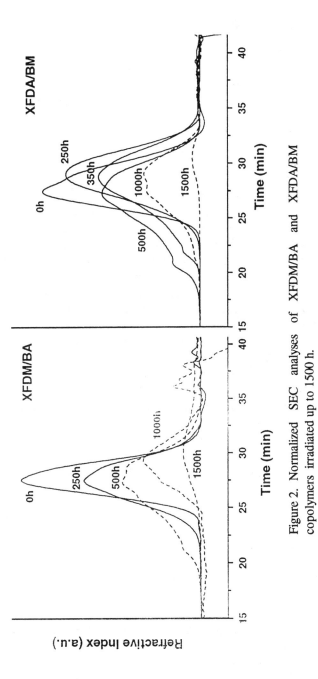

Figure 2. Normalized SEC analyses of XFDM/BA and XFDA/BM copolymers irradiated up to 1500 h.

broaden on the lower retention volume side, indicating the formation of molecules with molecular weights higher than the initial ones. At longer degradation times, on the other hand, the curves are only representative of the soluble fractions remaining in the samples.

In addition to the molecular weight distribution changes and the gel formation, the structural changes in the XFDM and XFDA copolymers are also very similar, and consist in the formation of hydroxyl-bearing groups through the typical mechanism of hydrocarbon oxidation (*11*), and of lactone groups, detectable from their characteristic absorption at 1785 cm^{-1}. The latter ones are invariably found as the photooxidation products of different types of structures like polyolefins (*12-14*) and acrylate homopolymers (*15*). The similarity in chemical transformations of XFDM/BA and XFDA/BM is reflected by the curves in Figure 3, where the increase of the infrared absorption at 1785 cm^{-1} is reported for these two samples.

The irradiation in air of TFEM/BVE and EM/BVE induces large changes in the samples, from both structural and molecular point of view. The main effects on the two copolymers are the development of oxidized structures (hydroperoxides, alcohols, lactones and acids), as detected from the FTIR spectra, and the extensive chain scissions, which may be visually appreciated by the displacement towards higher retention times of the SEC curves reported in Figure 4. The progress of these reactions is larger in the EM/BVE copolymer, and this may be clearly seen in the comparison of the numbers of chain scissions per initial polymer molecule, calculated from the values of number average molecular weights measured during degradation (16) and plotted in Figure 5.

Vinyl ether units in polymer samples have been reported to be quite sensitive to photooxidation (*17*), with development of new carbonyl containing functional groups and suggestions about reactions leading to homolysis of the backbone carbon–carbon bonds. The strong molecular weight decrease measured in our copolymer samples after the irradiation indicates that the same chain scission mechanism is present, which can be attributed to the effect of the vinyl ether units.

In fact, the photodegradative behaviour appears to be strongly affected by the presence of the vinyl ether units. As a matter of fact TFEM and EM homopolymers, subjected to the same irradiation conditions, undergo chain scissions to a much lesser extent, and very little or no oxidation of the macromolecular structure can be observed. On the other hand, oxidation in the vinyl ether copolymer with TFEM is lower than with EM, as it is shown by the respective increments of the lactone absorption, reported in Figure 6.

All copolymers of vinyl ethers with acrylic monomers fluorinated at, or close to, the vinyl group were generally characterized by a high degree of comonomer alternation. It was argued that such structural regularity, with fluorinated backbone C2 units alternating with backbone C2 units containing a labile tertiary hydrogen (e.g. acrylate or vinyl ether units), could stabilize the systems, inhibiting the backbone oxidation and the chain scission reactions. The first results on the MTFMA/EHVE copolymer showed a moderate effect of the CF$_3$ group. The molecular weight

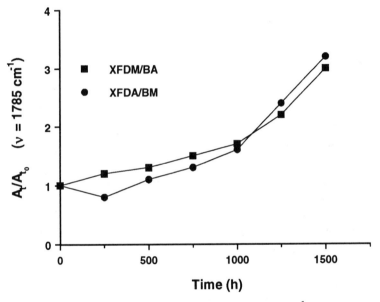

Figure 3. Change of the infrared absorption at 1785 cm^{-1} in XFDM/BA and XFDA/BM copolymers irradiated up to 1500 h.

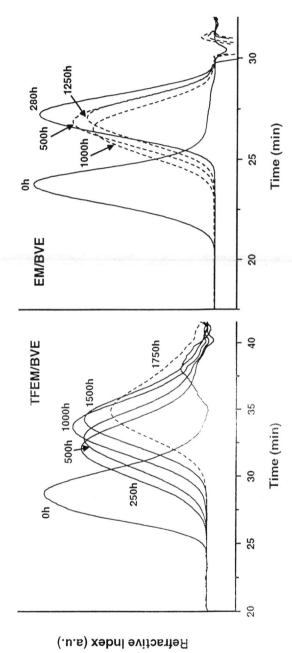

Figure 4. Normalized SEC analyses of irradiated copolymers of BVE with fluorinated (TFEM) and corresponding unfluorinated methacrylates.

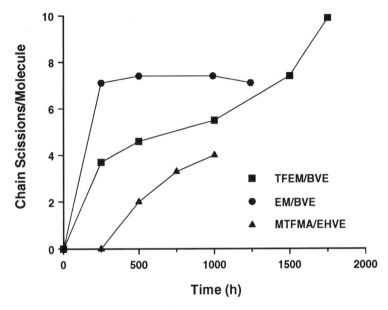

Figure 5. Numbers of chain scissions per molecule in irradiated TFEM/BVE, EM/BVE and MTFMA/EHVE copolymers.

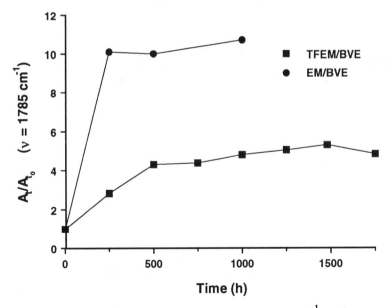

Figure 6. Change of the infrared absorption at 1785 cm^{-1} in irradiated copolymers of BVE with fluorinated (TFEM) and corresponding unfluorinated methacrylates.

distributions changes reported in Figure 7 can be compared with those of a side chain fluorinated copolymer (TFEM/BVE) and of its unfluorinated analogous (EM/BVE) in Figure 4. The smaller displacement of the curves corresponds to the lower number of chain scissions, which are plotted in Figure 5 together with scissions in TFEM/BVE and EM/BVE. The oxidation reactions in MTFMA/EHVE are similar to those already described for the other samples, indicating that in this case the F atoms are too far from the weak tertiary C-H bonds to allow a significant electronic effect for polymer stabilization.

The preliminary results obtained with the copolymers HFIFA/BVE and MFPE/BVE indicate that the photoinduced reactivity of the vinyl ether units is still significant, and polymer degradation is not suppressed. Structural modifications are probably controlled also in these samples by the reactions on the vinyl ether units, although the infrared spectra changes are quite different in comparison with the previous cases. Detailed evaluation of the photooxidative degradation reactions taking place in all the polymers is currently in progress.

In the case of the commercial Lumiflon LF200, very fast and complete crosslinking of the films was obtained by irradiating in air, as a result of the presence of reactive hydroxy-groups in the copolymer. However, the FT-IR spectra measured after more than 2000 h irradiation showed very small changes, indicating that the $-CF_2-CFCl-$ units alternating with vinyl ethers strongly inhibit the formation of oxidized structures. LF200 is therefore an interesting reference material for strictly alternating copolymers of vinyl ethers with comonomers bearing fluorine on the backbone carbon atoms. It is worthwhile to say that preliminary test on stone substrates have shown that the presence of vinyl ether units in acrylic based copolymers gave coating materials with good protective properties (Alessandrini, G.; Aglietto, M.; Castelvetro, V.; Ciardelli, F; Peruzzi, R.; Toniolo, L. *J. Appl. Polym. Sci., in press*), if compared with their fully acrylic analogues.

Conclusions

The results obtained so far, aiming at the development of new polymeric structures for the protection of valuable stone surfaces, allow to narrow the field of future investigations. Side chain fluorination of acrylic copolymers does improve the hydrophobicity in thin films, although its photostabilizing action is only moderate; in addition, structural and morphological order within the film can affect its water permeability and wettability (18) as much as its average chemical composition. Future efforts will therefore be focused on the more promising polymeric structures, i.e. those that join a good photochemical stability with an improved protection behavior as compared to conventional unfluorinated coating materials.

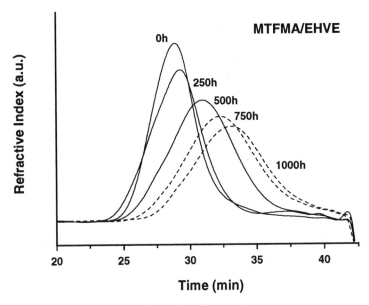

Figure 7. Normalized SEC analyses of MTFMA/EHVE copolymer irradiated up to 1000 h.

142

Acknowledgments

Financial support from the C.N.R. "Safeguard of Cultural Heritage" Target Project (http://soi.cnr.it/~tminfo/culturalheritage/) is gratefully acknowledged.

Literature Cited

(1) Ciardelli, F.; Aglietto, M.; Montagnini di Mirabello, L.; Passaglia, E.; Ruggeri, G. *Paints and Varnishes* **1996**,*72* (3), 21.

(2) Thomas, R. R.; Anton, D. R.; Graham, W. F.; Darmon, M. J.; Sauer, B. B.; Stika, K. M.; Schwartzfager, D. G. *Macromolecules* **1997**, *30*, 2883.

(3) Guo, H.-Q.; Kajiwara, A.; Morishima, Y.; Kamachi, M. *Polym. Adv. Techn.* **1997**, *8*, 196.

(4) Krupers, M.; Möller, M.; *Macromol. Chem. Phys.* **1997**, *198*, 2163.

(5) Park, I. J.; Lee, S.-B.; Choi, C. K. *Polymer* **1997**, *38*, 2523.

(6) Ciardelli, F.; Aglietto, M.; Montagnini di Mirabello, L.; Passaglia, E.; Giancristoforo, S.; Castelvetro, V.; Ruggeri, G. *Progr. Org. Coatings* **1997**, *32*, 43.

(7) Charola A. E.; Delgado Rodrigues J. *Science and Technology for Cultural Heritage* **1996**, *5*, 111.

(8) Puterman, M.; Jansen, B.; Kober, H. *J. Appl. Polym. Sci.* **1996**, *59*, 1237.

(9) "Normal" Protocol 20/85. "Conservation Works: Planning, Execution and Preventive Evaluation" ICR-CNR Ed., Rome (Italy), **1985**.

(10) Castelvetro, V.; Aglietto, M.; Montagnini di Mirabello, L.; Toniolo, L.; Peruzzi, R.; Chiantore, O. *Surf. Coating Internat.* **1998**, *81*, 551.

(11) *Degradation and Stabilisation of Polyolefins*; Allen, N. S., Ed.; Applied Polymer Science Publisher Ltd.: London, 1983.

(12) Adams, J. H. *J. Polym. Sci. A-1* **1970**, *8*, 1077.

(13) Ginhac, J. M.; Gardette J. L.; Arnaud R.; Lemaire J. *Makromol. Chem.* **1981**, *182*, 1017.

(14) Tidjani A.; Arnaud. R. *Polym. Degrad. Stab.* **1993**, *39*, 285.

(15) Chiantore O.; Trossarelli L.; Lazzari M. *Polymer* **2000**, *41*, 1657.

(16) Grassie, N. in *Degradation in Polymer Science*; North Holland: Amsterdam; Vol. 2, p. 1443.

(17) Gardette, J.L.; Sabel, H.D.; Lemaire, J. *Angew. Makromol. Chem.* **1991**, *188*, 113.

(18) Tirelli, N.; Ahumada, O.; Suter, U.W.; Menzel, H.; Castelvetro, V. *Macromol. Chem. Phys.*, **1998**, *199*, 2425

Chapter 11

Fluorinated Polymer Claddings for Optical Fibers

Arturo Hale, Kathryn W. Quoi, and David J. DiGiovanni

Bell Laboratories, Lucent Technologies, 700 Mountain Avenue, Room 7E–217, Murray Hill, NJ 07974

The standard process to coat glass optical fibers with polymers consists of applying a liquid coating with a die and UV-curing it. It is not straightforward to employ this procedure with low-index fluorinated coatings because of their very low viscosity. This work shows that the viscosity of fluorinated diacrylates can be increased by two orders of magnitude by controlled crosslinking with a tetrafunctional silane. The resulting material has a viscosity that is high enough for a fiber coating operation. Photocured coatings with very low refractive index (1.323) have been obtained and employed as claddings for glass optical fibers. Losses as low as 3 dB/km were measured on these fibers.

Optical fibers consist of a transparent core surrounded by an also transparent cladding layer whose index of refraction is lower than that of the core. Light can be guided through the core by total internal reflection. Optical fibers used for tele-communications typically have a germanium-doped silica core, a silica cladding, and a protective polymeric coating that usually plays no optical role. Despite their higher optical losses, there are some specific instances in which it is advantageous to use organic polymeric materials instead of silica glass as optical claddings. Use of a polymeric cladding enables the construction of thicker fibers while maintaining flexibility, a useful feature that can lower the cost of connectorization. In addition, the refractive index of polymers can be made much lower than that achievable by doping silica with fluorine or boron. This lower refractive index provides a larger numerical aperture (N.A.), thereby increasing the acceptance angle for the incoming light to be guided through the fiber. This is of great importance in applications such as fiber lasers, which can be used for telecommunications, printers, material processing, and medical applications.

A fiber laser typically comprises an optical fiber whose core is doped with a rare-earth element. Rare-earth elements have absorption and emission bands in the near-infrared region, and the lifetimes of the excited states are appropriate for building laser cavities at certain specific wavelengths. Erbium in particular is widely utilized today in the telecommunications industry as a dopant in optical fiber amplifiers at 1.55 μm; however, the popularity of erbium-doped fiber amplifiers (EDFA) has spurred interest in higher powers than those achievable with commercial devices. The output of EDFAs is limited only by the amount of single-mode coupled pump power that one has available for pumping into one of the many absorptions of erbium-doped fibers (1). The optical powers that are currently available from single-mode fiber-pigtailed laser diodes are limited by the small spot size required for single mode operation and by the intrinsic materials properties of the laser diode themselves, such as facet damage (1,2).

A clever solution to this limitation is to break the constraint on single-mode diode coupling by using a double-clad laser, also known as cladding-pumped fiber laser (CPFL) (2), shown schematically in Figure 1. This device consists of a silica core doped with a rare-earth element, a silica cladding, and an outer polymeric cladding whose refractive index is much smaller than that of the silica cladding. Thus, the entire inner cladding region becomes a waveguide itself. When pump light from a high power multimode laser diode is introduced into the glass cladding, it propagates along the fiber, occasionally crossing the core, where it is absorbed by the rare-earth dopant that can be made to lase. The fiber is non-circular to allow all the pump rays to eventually cross the core; otherwise, much of the light would follow helical paths and would not get absorbed (2).

To collect the greatest amount of pump light, it is necessary to raise the numerical aperture of the cladding pumped fiber laser by lowering the refractive index of the polymer outer cladding as much as possible. The polymer with the lowest refractive index currently available is Teflon AF, an amorphous copolymer of (perfluoro-2,2-dimethyldioxole and polytetrafluoroethylene) made by DuPont. It is not practical, however, to make optical fiber using Teflon AF as a cladding because of manufacturability issues. In order to understand these issues better it is necessary to describe the process currently employed to manufacture optical fibers.

This process starts by making a glass rod (known as the "preform") that contains the basic structure of the glass optical fiber, i.e. a doped silica core and a cladding, by a vapor deposition process (3,4). This preform, which has a diameter on the order of centimeters, is then made into an optical fiber by the process depicted in Figure 2. The preform is placed inside an induction furnace, where it is heated to about 2000°C. The preform softens and is drawn into a thin fiber (the diameter is typically 125 μm, but it can be different for other applications). This thin glass fiber cools down as it travels down the draw tower; because it is freshly drawn from the melt, its surface is pristine. To protect the delicate glass surface, it is necessary to coat it with a protective layer before any solid comes in contact with it. The current technology involves coating the fiber on-line as it is being drawn by passing it through a coating

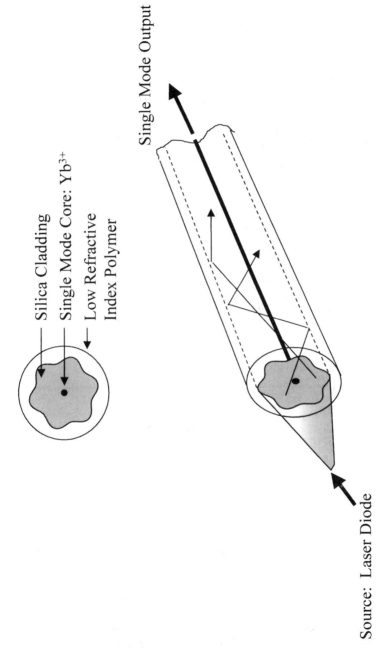

Single Mode Output

Silica Cladding

Single Mode Core: Yb^{3+}

Low Refractive Index Polymer

Source: Laser Diode

Figure 1. Schematic of a cladding-pumped fiber laser.

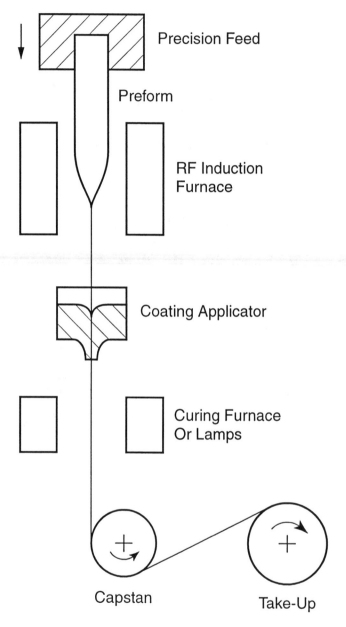

Precision Feed

Preform

RF Induction
Furnace

Coating Applicator

Curing Furnace
Or Lamps

Capstan

Take-Up

Figure 3. Calculated weight-average molecular weight as a function of stoichiometric imbalance at full conversion of the limiting reagent.

die where a liquid pre-polymer is applied. The pre-polymer is immediately cured by UV-induced polymerization. The fiber with its protective polymer coating can then be taken up on a spool. This polymer coating is only for mechanical protection; it plays no optical role.

The adaptation of this technology to produce optical fibers with low index polymer claddings is not straightforward. If one wants to use a thermoplastic such as Teflon AF as a cladding, one would need to either extrude the polymer onto the fiber or to apply it from solution. In the case of extrusion, it is very difficult to obtain a fiber with adequate mechanical strength, because of damage on the glass introduced by particles present in the polymer or introduced by the extrusion process. If one applies the polymer coating from a solution, it is possible to remove particulates by filtration, but the process is necessarily slow because of the need for solvent removal; in addition, one can only apply a thin layer at a time, which may not be enough to protect the fiber from mechanical damage during processing.

To manufacture polymer-clad optical fibers using currently available fiber production technologies, it would be desirable to employ a UV-curable material as a cladding. The most widely-used UV-curable coating materials are based on acrylates, because of their extremely fast curing rates. However, most commercially available acrylates have indices of refraction higher than that of silica, and therefore they cannot be used as claddings. In order to obtain refractive indices low enough to be attractive for fiber laser applications, one needs to use fluorinated coatings. There are several fluorinated acrylates commercially available; however, they cannot be used directly in the draw process shown in Figure 2, because their viscosities are much too low. If one attempts to coat a glass fiber with a low viscosity coating, surface tension will cause the coating to form beads before it cures (5), leaving portions of the fiber unprotected.

It has been shown that the viscosity of the pre-polymer should be at least about 3,000 cP to produce a uniform coating with a thickness on the order of tens of microns (5). As a point for comparison, L-9367, a diacrylated perfluoropolyether available from 3M (shown below) has a viscosity of only 35 cP.

$$CH_2=CHC(O)OCH_2-(C_2F_4O)_m-(CF_2O)_nCH_2OC(O)CH=CH_2 \qquad MW=2,000$$

L9367

One approach for obtaining a low index UV-curable coating with an viscosity high enough for die coating is to blend a fluorinated polymer with UV-curable fluorinated monomers (6,7). However, it is somewhat difficult to obtain the right composition that will not phase-separate and induce light scattering upon polymerization of the monomers. The lowest refractive index reported using this approach is 1.349 (6), but more common values are on the order of 1.38 (6,7). An alternative procedure that provides lower refractive indices is presented here. This approach consists of pre-reacting a di-functional fluorinated acrylate (such as L-9367) with a multifunctional crosslinker to create a UV-reactive high molecular weight branched structure with a high viscosity (8).

Theory

The viscosity of a polymer depends on two factors: a structure factor and a friction factor (9). The structure factor is related to the size of the molecule: a larger molecule presents larger resistance to flow than a smaller molecule, leading to a higher viscosity. The friction factor is related to the intermolecular interactions between the molecules as they flow past each other as well as to their flexibility.

The friction factor of highly fluorinated organic molecules is inherently low because of the absence of strong intermolecular interactions; therefore, the obvious approach to increasing the viscosity of a fluorinated composition is to increase the size of the molecules. At the same time, it is necessary to ensure that these larger molecules are still UV-curable. A convenient way to achieve this objective is to increase the molecular size by a three-dimensional branching reaction. This permits the formation of very large molecules with many reactive (UV-curable) groups (one at the end of each branch).

If a branching reaction is not properly controlled, it can lead to the formation of a three-dimensional network (gelation). For this application it is necessary to prevent gelation at this stage, because the objective is to produce a higher viscosity UV-curable material that is still liquid. This is achieved by employing a step-polymerization for the branching process, instead of the harder to control chain mechanism typically associated with acrylate polymerization.

The statistics of network formation during step polymerization have been described by many authors (10,12). For the step co-polymerization of a molecule containing f reactive groups A with another molecule containing g reactive groups B, the condition at which this system gels is given by (12):

$$r\alpha^2 = 1/[(f-1)(g-1)] \qquad (1)$$

where r, the stoichiometric imbalance, is the ratio of the number of moles of A reactive groups to B reactive groups, α is the extent of reaction or conversion of A groups, f is the functionality of the molecule containing A groups, and g is the functionality of the molecule containing B groups. This formula assumes that all groups have the same reactivity, that they react independently of each other, and there is no cycle formation.

For a given set of functionalities, the above formula can be used to calculate the critical ratio of the two components needed to gel the mixture if the limiting reagent is allowed to react to completion. If A is the limiting reagent, then $\alpha = 1$, and the critical gelation ratio is given by

$$r_c = 1/[(f-1)(g-1)] \qquad (2)$$

In order to obtain a high molecular weight branched structure starting from a difunctional acrylate (g = 2), it is necessary to co-polymerize it with a molecule whose functionality is at least three. In the case of a step co-polymerization with a tetrafunctional crosslinker (f = 4), r_c = 1/3, that is, a mixture that has three times as many B groups (acrylate groups) as A groups will gel upon full conversion of the A groups. In order to prevent gelation, r_c should be lower than 1/3. The actual ratio needed to obtain the necessary increase in viscosity can be estimated from molecular weight calculations, because the viscosity depends on molecular weight. In fact, Hale and Macosko (*13*) have shown that the viscosity of a mixture of multifunctional monomers undergoing a branching step polymerization is directly proportional to the weight-average molecular weight.

In order to calculate the weight-average molecular weight of a difunctional molecule co-polymerized with a tetrafunctional crosslinker via a step mechanism, one can use a recursive approach proposed by Macosko and Miller (*12*). Using this procedure one can generate Figure 3, which shows the calculated weight-average molecular weight (normalized to the initial molecular weight of the mixture) as a function of normalized stoichiometric imbalance of a difunctional molecule co-polymerized with a limited amount of a tetrafunctional crosslinker. Each point corresponds to a system where the tetrafunctional crosslinker has reacted to completion, but with plenty of unreacted groups belonging to the difunctional molecule because the latter is present in excess; the abscissa represents the stoichiometric imbalance (the ratio of reactive groups present initially in the tetrafunctional crosslinker to reactive groups originally present in the difunctional molecule) normalized to the critical stoichiometric ratio r_c. As the stoichiometric imbalance approaches the critical ratio r_c, the molecular weight increases dramatically, eventually diverging at the gel point (r = r_c).

Because the viscosity is proportional to the molecular weight as discussed previously, Figure 3 also represents the normalized viscosity of these mixtures as a function of stoichiometric imbalance. As was explained in the introduction, the purpose of this work is to obtain a coating with a viscosity of around 3,000 cP, which is about 85 times higher than the viscosity of the starting diacrylate. Using Figure 3, one can see that it is possible to increase the viscosity of the initial diacrylate by a factor of 85 if one co-polymerizes the diacrylate with about 99% of the amount of tetrafunctional species needed to gel the material.

Chemistry

In order to apply the aforementioned ideas to increasing the viscosity of a diacrylate such as L-9367, one needs a multifunctional molecule capable of step-wise reaction with acrylates. There are several well-known step-wise reactions involving acrylates. Examples include addition to silicon hydrides (hydrosilylation), Michael addition to thiols, and Michael addition to amines. Hydrosilylation is particularly attractive for low refractive index coatings, because organosilicon molecules tend to have lower refractive indices than amines or thiols.

Hydrosilylation reactions encompass the addition of a silicon hydride to an ethylenic unsaturation; catalysts based on transition metals such as platinum enable these reactions to be carried out at moderate temperatures   The hydrosilylation of

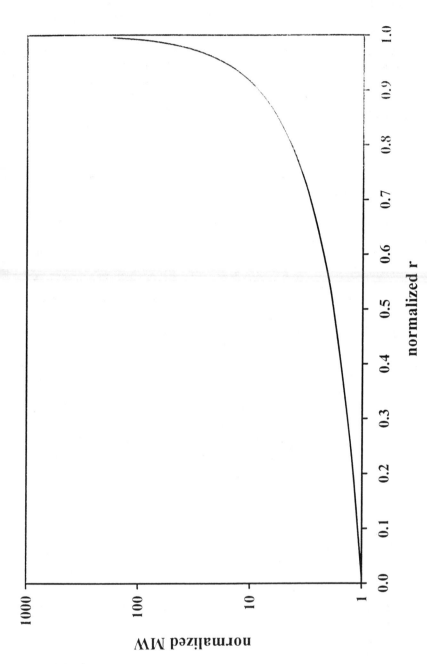

Figure 2. Schematic diagram of an optical fiber drawing and coating process.

most types of vinyl groups proceeds cleanly through the addition of the silicon atom to the terminal carbon of the vinyl. However, hydrosilylation of unsaturated esters such as acrylates can proceed in three different ways: α addition, β addition or 1,4 addition, as shown in Scheme 1 (14-19).

The relative importance of each reaction depends on the particular reagents used, the reaction temperature and the catalyst. The influence of the groups attached to the silicon atom is especially important. Ojima et al. (14) have found that both phenyldimethyl silane and ethyldimethyl silane yield 100% β product on methyl acrylate and ethyl acrylate, while dimethylchlorosilane yields 100% α product. Triethylsilane gives a mixture of β and a disilylated product. Speier et al. (15) found both α and β products in the Pt/C catalyzed reaction of methyl acrylate with methyldichlorosilane. Belfield and coworkers (16) found all three products (α ,β, and 1,4) in the Pt catalyzed reaction of methyl acrylate with 1,3,5,7-tetramethylcyclotetrasiloxane in the ratio 2.2:1.8:1, respectively. The presence of bulky groups on the acrylate also affects the reaction path. For example, Belfield et al. (16) showed that acrylates with large, bulky ester functionalities (e.g. 2,6-di-*t* -butyl-4-hydroxymethylphenyl acrylate) favor β addition. Methacrylates have also been reported to favor β addition with platinum catalysts (16-19). Temperature can also alter the product distribution, as shown by Skoda-Földes et al. (18), who reported that α addition is preferred at low temperature, while β addition is favored at high temperature on a methyl acrylate/triethyl silane system. There are some catalysts that can promote the formation of certain products. For example, Boudjouk et al. (19) have reported that acrylates can be exclusively β-hydrosilylated with methyldichlorosilane in the presence of copper salts and tetramethylenediamine. When this reaction is carried out in the presence of platinum instead, a mixture of α and β products is obtained (15).

For the purpose of this work, tetrafunctional silanes were employed as crosslinkers for the diacrylate. The silanes selected were tetrakis(dimethyl-siloxysilane) and 1,3,5,7-tetramethylcyclotetrasiloxane. The latter was shown to yield a mixture of α, β and 1,4 products upon reacting with acrylates, as explained above (16). Although all three reactions can be used to increase the viscosity of an acrylate as described previously, the β addition is the most desirable, since it yields the most stable product. The α addition product contains the Si–C–C=O group which is known to give carbon-silicon cleavage with acids and bases (20), while the 1,4 product can be decomposed by methanolysis (16). In any event, as long as the β addition is significant, the presence of other reactions should not be a problem, since the material will be subsequently crosslinked by acrylate polymerization, which will provide integrity to the coating.

Thus, by reacting the diacrylate L-9367 with an amount of polyfunctional silane just below the critical amount needed for gelation, a higher viscosity material can be obtained. Since there is an excess of acrylate groups, the viscosity-increased polymer still has unreacted acrylates available for the subsequent UV-polymerization during the fiber drawing process.

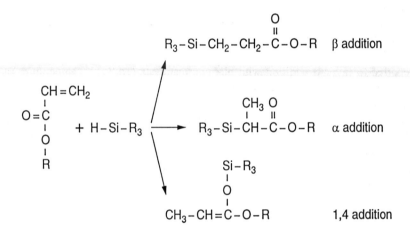

Scheme 1

Experimental

Materials.

L-9367 (a diacrylated perfluoropolyether), L-9186 (pentadecafluorooctyl acrylate) and FX-189 [2-(N-butylperfluorooctanesulfonamido) ethyl acrylate] were obtained from 3M. Trimethoxysilane, tetrakis(dimethylsiloxy)silane (TDMSS) and tetramethylcyclotetrasiloxane (TMCTS) were purchased from Gelest Inc., and the platinum catalyst (platinum divinyltetramethyldisiloxane complex) is available in solution (2-3% in xylene) as "PC072" from United Chemical Technologies. The photoinitiator used was Darocur 1173 from Ciba Specialty Chemicals.

Instrumentation

NMR Spectroscopy.
Proton NMR spectra were obtained using a Bruker AM360 spectrometer.

Infrared Spectroscopy.
Infrared spectra were acquired on a Nicolet Magna IR 760 Fourier transform spectrometer using an MCT-A detector.

Viscosity Measurements.
Viscosities were measured at room temperature in a Brookfield viscometer at 10-50 rpm using a Couette geometry on a small sample (2 cc) cell.

Refractive Index Measurements.
Refractive indices were measured on flat, free standing films using a Metricon prism coupler. The values reported were taken at room temperature using a HeNe laser (633 nm).

Fiber Drawing.
Polymer-clad glass fiber was drawn on a draw tower represented schematically in Figure 2. A quartz rod was used as precursor for the glass fiber; the product consisted of a 125 μm diameter silica core surrounded by a 250 μm diameter fluorinated polymeric cladding.

Optical Measurements.
Spectral loss was measured using the cutback technique on a 50-meter length of 125-micron-diameter silica core polymer-clad sample. Test fibers were loosely coiled on a

tabletop to eliminate excess microbending loss caused by overwinding of spooled fiber. Because loss for such high N.A., large core, multi-mode guides is in general both launch- and fiber-length- dependent, a variable launch test set capable of independently controlling launch N.A. and spot size was used to first determine launch conditions resulting in an equilibrium-mode distribution over a short 2-meter length of fiber. A measured N.A. of 0.33 as calculated from the far-field mode patterns of both long and short lengths of test fiber corresponded to the steady-state launch condition that was used to measure loss. This technique enables the measurement of signal loss for arbitrary lengths of low index polymer-clad fibers.

Procedure and Results

In order to confirm that hydrosilylation indeed proceeds as expected on our fluorinated acrylates, infrared spectroscopy and proton NMR spectroscopy were performed on model compounds. A 2:1 mole mixture of pentadecafluorooctyl acrylate and trimethoxysilane was reacted for 15 hours under nitrogen at 70°C using 0.5 wt. % of PC072 platinum solution as catalyst. The reaction of the Si-H group was verified by the disappearance of the 2130 cm^{-1} infrared band. Proton NMR spectroscopy revealed a significant amount of β addition product (2.4, 1.2 ppm), as well as some 1,4 addition (3.4 ppm) and a small amount of α addition (1.45 ppm) products.

As mentioned before, the β addition is the most desirable, because it yields the most stable product. Nevertheless, all three reactions contribute to raising the viscosity, and the species that promote degradation of the α and the 1,4 products are not present in the current formulations.

In order to study the viscosity behavior of stepwise-formed networks of diacrylate with tetrafunctional silanes, L-9367 was mixed with different ratios of TDMSS in separate vials, and the hydrosilylation reaction was carried out at 75°C for 15 hours using 0.3% by weight of PC072 platinum solution as catalyst. The samples where the weight ratio of TDMSS to L9367 exceeded 0.03 gelled. This is consistent with theory, which predicts that the critical gelation ratio is about 0.027 g TDMSS/g L9367 assuming that the molecular weight of L-9367 is 2,000 and that it is 100% difunctional (nominal values). The viscosity of the ungelled samples was then measured in the Brookfield viscometer. The results are shown in Figure 4, where the stoichiometric ratio in the abscissa represents the initial ratio of moles of SiH to moles of acrylate groups. The continuous line represents a prediction based on the calculated molecular weights shown in Figure 3. As mentioned earlier, the theoretical gel point occurs at a stoichiometric imbalance of 1/3. The agreement between the prediction and the experimental results is quite good.

A somewhat lower refractive index can be obtained by employing TMCTS instead of TDMSS. A mixture containing L9367 and 0.093 wt.% of PC072 platinum solution was reacted with 0.027 g TMCTS / g L9367 (this proportion of reagents is close to the critical gelation ratio). The mixture was held at 75 - 80°C for three days,

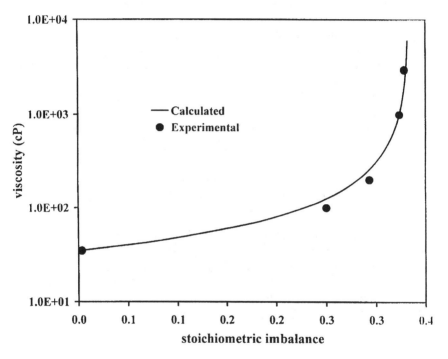

Figure 4. Viscosity as a function of stoichiometric imbalance for mixtures of TDMSS fully reacted with different excess amounts of L-9367.

after which the hydrosilylation reaction was essentially complete as determined by the absence of the Si-H infrared band at 2130 cm^{-1}. The viscosity of this mixture at room temperature was 2,700 cP, high enough to be employed as a coating for an optical fiber. This mixture was blended with 0.9 wt.% Darocur 1173 (photoinitiator) and 1 wt. % of 2-(N-butylperfluorooctanesulfonamido) ethyl acrylate (available from 3M as FX-189, used to help compatibilize the photoinitiator) to obtain a photocurable formulation.

The photocurable formulation was cast between two glass slides using a 10 mil thick FEP frame as a spacer, and cured with a high intensity mercury lamp. After removing this film from the glass slides, its refractive index was found to be 1.323 as measured by a prism coupler at 23°C and 633 nm. This is one of the lowest, if not the lowest, refractive indices reported for a UV-cured polymer suitable for cladding optical fibers.

Several hundreds of meters of optical fiber were drawn using this material as cladding following the procedure illustrated in Figure 2. The viscosity of the liquid coating was high enough to allow a smooth drawing operation. The optical quality of the resulting fiber was very good, as shown by the spectral loss measurements (Fig. 5). Optical attenuations as low as 3 dB/km were recorded at a wavelength of 1.1 μm. The peak at 0.945 μm corresponds to Si-OH in the glass.

Conclusions

It has been shown that the viscosity of fluorinated diacrylates can be increased by two orders of magnitude by controlled crosslinking with a tetrafunctional silane. The resulting material has a viscosity that is high enough for a fiber coating operation. Photocured coatings with very low refractive index (1.323) have been obtained and employed as claddings for optical fibers. Losses as low as 3 dB/km at 1.1 μm were measured on these fibers.

Acknowledgments

The authors thank Susan Stein for acquiring the NMR spectra.

Literature Cited

1.   White, A. E.; Grubb, S. G. In *Optical Fiber Telecommunications IIIB*; Kaminow, I. P.; Koch, T. L., Eds.; Academic Press: San Diego, CA, 1997; pp. 267-318.

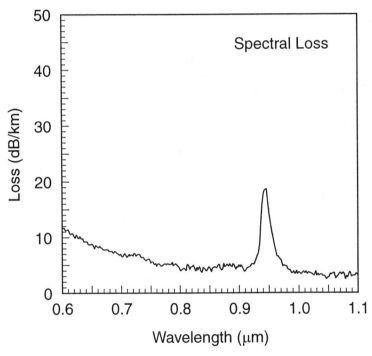

Figure 5. Measured spectral loss on a glass fiber coated with a UV-cured fluorinated polymer cladding.

158

2. DiGiovanni, D. J.; Jablonowki, D. P.; Yan, M. F. In *Optical Fiber Telecommunications IIIA;* Kaminow, I. P.; Koch, T. L., Eds.; Academic Press: San Diego, CA, 1997; pp. 63-91.

3. French, W. G.; Jaeger, R. E.; MacChesney, J. B.; Nagel, S. R.; Nassau, K.; Pearson, A. D. In *Optical Fiber Telecommunications*; Miller, S. E.; Chynoweth, A. G., Eds.; Academic Press: San Diego, CA, 1979; pp. 233-261.

4. Nagel, S. R. In *Optical Fiber Telecommunications II;* Miller, S. E.; Kaminow, I. P., Eds.; Academic Press, San Diego, CA, 1988; pp. 121-215.

5. Blyler, L. L.; DiMarcello, F. V. In *Encyclopedia of Physical Science and Technology;* Academic Press: San Diego, CA, 1987, Vol. 9; p. 647.

6. Hashimoto, Y.; Kamei, M.; Umaba, T. *U.S. Patent 4,786,658.*

7. Minns, R. A.; Bloom, I. B. K.; Ramharack, R. *U.S. Patent 5,024,507.*

8. Hale, A. *U.S. patent 5,756,209.*

9. Berry, G. C.; Fox, T. G. *Adv. Polym. Sci.* **1968,** *5*, 261.

10. Flory, P. J. *Principles of Polymer Chemistry;* Cornell University Press: Ithaca, NY, 1953, pp. 347-398.

11. Stockmayer, W. J. *J. Chem. Phys.* **1943,** *11*, 45.

12. Macosko, C. W.; Miller, D. R. *Macromolecules* **1976,** *9*, 199.

13. Hale, A; Macosko, C. W. *Poly. Mat. Sci. Eng.* **1988,** *59*, 1196.

14. Ojima, I.; Kumagai, M.; Nagai, Y. *J. Organomet. Chem.* **1976,** *111*, 43.

15. Speier, J. L.; Webster, J. A.; Barnes, G. H. *J. Am. Chem. Soc.* **1957,** *79*, 974.

16. Belfield, K. D.; Lin, X. Z.; Cabasso, I. *J. Polym. Sci. Part A: Polym. Chem.* **1991,** *29*, 1073.

17. Curry, J. W.; Harrison, G. W. *J. Org. Chem.* **1958,** *23*, 627.

18. Skoda-Földes, R; Kollár, L.; Heil, B. *J. Organomet. Chem.* **1991,** *408*, 297.

19. Boudjouk, P.; Kloos, S.; Rajkumar, A. B. *J. Organomet. Chem.* **1993,** *443*, C41.

20. Sommer, L. H.; MacKay, F. P.; Steward, O. W.; Campbell, P. G. *J. Am. Chem. Soc.* **1957,** *79*, 2764.

Chapter 12

Ultra Water Repellent Thin Films Prepared by dc Plasma Polymerization of Vinylidene Fluoride

Daisuke Sato, Mitsutoshi Jikei, and Masa-aki Kakimoto*

Department of Organic and Polymeric Materials, Tokyo Institute
of Technology, Meguro-ku, Tokyo 152, Japan

Thin films of plasma polymerized vinylidene fluoride (PPVF) were
produced by D. C. glow discharge methods. The effects of
experimental parameters such as atomic content and surface roughness
are discussed. PPVF films prepared with varying conditions such as
the applied voltage, gas pressure, and selection of electrodes had almost
identical chemical content as evidenced by IR spectroscopy and ESCA
measurements. The contact angles varied from 83 to 162° on these
film surfaces, where the values were higher in the films prepared on the
anode. Atomic force microscopy showed a clear relationship between
the surface roughness, which varied from 2 to 270 nm, and the measured
contact angles. A higher value of contact angle correlates to a rougher
film surface.

Plasma polymerization of vaporized organic compounds is a superb method to
produce pinhole free thin films. In previous papers, we reported on D. C. glow
discharge techniques for a plasma polymerization of both hydrogenated and
fluorinated aromatic compounds (1-4). It was found that the resulting thin films
derived from the hydrogenated aromatic compounds consisted of only carbon,
whereas both carbon and fluorine atoms were detected in films prepared from the
fluorinated aromatic precursors. These results indicated that the hydrogen atom
does not get incorporated into the plasma polymerized films, but fluorine remains in
the films under the high power plasma polymerization conditions. In these previous
studies, contact angles of water against the surface of the plasma polymerized films
were 70° and 100° for films formed from non-fluorinated and fluorinated aromatic
precursors, respectively. Control of conditions such as applied voltage and gas
pressure is easier in D.C. plasma polymerization compared to the usual method
which uses radio frequency (RF). Furthermore, using the D. C. method, films with
different sturctures can be prepared on the cathode and the anode.

One application of the plasma polymerization technique is the formation of water repellant thin films.   Hozumi et al. have reported the preparation of water repellent films using RF and microwave plasma enhanced CVD methods starting from tetramethylsilane and fluoro alkyl silanes (5-9).   In our previous work, D. C. plasma polymerization of aromatic fluorocarbons did not afford ultra water repellent thin films.   The ability of a film to repel water is reflected in the contact angle of water on the film surface.   A value of 160° is taken as ultra water repellency.

In this paper, D.C. plasma polymerization of vinylidene fluoride (VF) to produce the ultra water repellent films is described.   VF is an inexpensive fluorinated gas, and few papers have reported its plasma polymerization (10).

Experimental
Plasma polymerization of vinylidene fluoride (VF).   The plasma polymerized vinylidene fluoride (PPVF) films were prepared by a D.C. glow plasma reactor, as shown in Figure 1, where the disks of anode and cathode electrode are arranged parallel to one another. In preparing PPVF film, the substrates such as glass and silicon plates were placed on the cathode or anode.

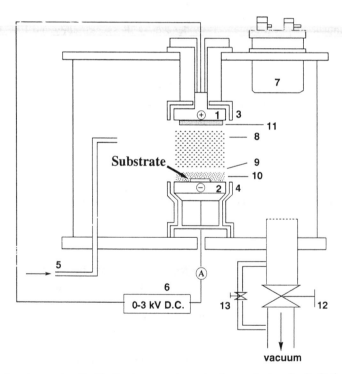

Figure 1.   Apparatus for D. C. plasma polymerization.   1. Anode, 2. Cathode,   3. and 4. Leakage discharge prevention cylinder, 5. Gas inlet for gaseous monomer, 6. High voltage power supply, 7. Sublimation chamber, 8. Positive column 9. Faraday dark space, 10. Negative glow phase, 11. Positive glow phase, 12. Main valve, 13 . Adjusting valve

In a typical experiment, the reactor was evacuated to below 10^{-2} Pa, before VF gas was introduced into the reactor from the gas cylinder at a flow rate of 10 mL/ min. While the pressure in the reactor was regulated between 10 and 30 Pa, a DC voltage from 0.5 to 1.5 kV was applied for an appropriate time to form plasma polymerized films on the anode or cathode. After the plasma reaction, the vacuum chamber was evacuated once again below 10^{-2} Pa for 5 minutes to minimize post reaction of the film surface with water and oxygen in air, and then the system was vented to atmospheric pressure.

Analysis.   A needle probe analyzer, DEKTAK3, was used for determination of the film thickness.   The contact angle of water was measured using the sessile drop method by a contact angle meter, Kyowa Interface Science Co., Model CA-A. Reflection absorption infrared spectra (RAS) at a resolution of 4 cm^{-1} were measured with   a   JEOL   JIR-MICRO   6000   equipped   with   a   nitrogen-cooled mercury-cadmium-telluride (MCT) detector.   A glass plate coated with ca. 100 nm of vacuum evaporated silver was used as a substrate for RAS measurements. Atomic constitution and concentration were measured by X-ray photoelectron spectroscopy (XPS) using a ULVAC-PHI-5500MT system. The spectra were acquired using monochromated Al Kα (1486.7 eV) radiation at 14 kV and 200 W. Measurements of the bulk constitution of PPVF films were obtained by bombardment of the films with argon ions at 3.0 kV and 5∼25 mA.   Atomic force microscopy (AFM) measurements were made using a Seiko Instruments SPA300 microscope. Mica cleaved by the scotch tape was used as a substrate for AFM experiments.

Results and discussion
Figure 2 shows the film thickness of PPVF films as a function of time grown with an applied voltage of 1.0 kV, a gas pressure of 30 Pa, and with a gas flow rate of 10 ml/min. The initial growth rate of film thickness of about 4 nm/sec gradually decreased to about 2.5 nm/sec after time.   The reason a decrease in film growth rate is observed is because a nonconductive coating grew on the electrodes during the plasma polymerization.   This phenomenon is usually observed in D. C. glow plasma polymerization.

The FT-IR reflection absorption spectra (RAS) of PPVF films that were grown on the cathode at 30Pa with different applied voltages are shown in Figure 3. Two major peaks are found around 1160 and 1260 cm^{-1}, and are attributed to carbon-fluorine stretches.   Small broad peaks from 1600 to 1800 cm^{-1} are attributed to carbonyls and carbon-carbon double bonds.   The three spectra have the same absorption features, indicating similar chemical structures are present in all three plasma polymerized films.   Furthermore, PPVF films prepared on the anode afforded the same FT-IR spectra as those grown on the cathode.   Thus, IR results show that films of PPVF prepared under these different experimental conditions have similar chemical structures.   XPS survey spectra of PPVF films grown at 1.0 kV and at 30 Pa on the both cathode and anode indicated that only carbon, fluorine and oxygen exist at the surface.   Because oxygen was not detected after sputtering of the film surface by argon ions, it is concluded that oxygen existed only at the film surface.

It is assumed that water and oxygen in the atmosphere reacted with active

162

species, and remained on the PPVF film surface after the plasma polymerization. Table 1 presents the atomic concentration of carbon, fluorine and oxygen for the films prepared at either 0.5 kV or 1.0 kV with the operating pressure of 30 Pa. Although the fluorine concentration is slightly higher in the films prepared on the cathode, the ratio of elements is not a big difference in each film.

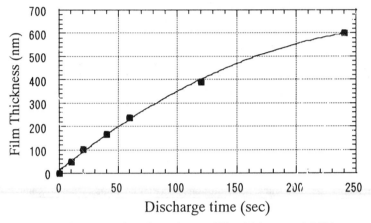

Figure 2.   Time dependence of film thickness at 1.0 kV.

Table 1.   Atomic concentration of consistent element of PPVF on the surface of films[a]

	Cathode		Anode	
	0.5 kV	1.0 kV	0.5 kV	1.0 kV
Carbon (%)	61.2	62.7	63.1	65.3
Fluorine (%)	35.1	34.0	31.5	31.6
Oxygen (%)	3.7	3.3	5.4	3.1

a) Gas pressure was 30 Pa, and gas flow rate was 10 ml/min

Figure 4 indicates change of the contact angles of water at the surface of PPVF films prepared on the glass substrate put on the cathode. The contact angles are almost the same regardless of the applied voltages at 10 Pa. Change of the contact angles with increase of the gas pressure depended on the applied voltage. Although the contact angle remained near $85°$ in the case of 0.5 kV, the value reached $120°$ at the applied voltage of 1.0 kV. The contact angles of PPVF films prepared on the cathode as well as anode are shown in Table 2. PPVF films prepared on the anode had larger contact angles than ones on the cathode, even though the fluorine content was less in the films prepared on the anode. The maximum contact angle on the anode reached $162°$ for the condition of   1.0 kV and 30 Pa.

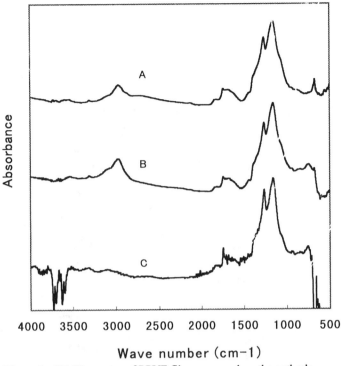

Wave number (cm⁻1)

Figure 3   FT-IR spectra of PPVF films prepared on the cathode.
Conditions: A) 1.0 kV, B) 0.8 kV, and C) 0.5 kV.   Gas pressure
is 30 Pa and flow rate is 10 ml/min in all cases.

Table 2.   Contact angles (degree) of water on PPVF films prepared in various
conditions

	Applied Voltage / Gas Pressure	0.5 kV	0.8 kV	1.0kV
Cathode	10 Pa	86	83	84
	30 Pa	90	108	119
Anode	10 Pa	86	85	87
	30 Pa	88	116	162

Figure 5 shows the AFM images of PPVF films deposited at 1.0 kV on the
cathode at different gas pressures.   Roughness of the surface was about 2 and 90 nm
in the cases of 10 and 30 Pa deposition gas pressures, whereas the contact angles
were 85 and 119 degrees, respectively.   The same tendency was observed in the

Figure 5. AFM images (scan area 1X1 μm, depth profile is indicated below the images) of PPVF films prepared on the cathode. A) Applied voltage is 1.0 kV and Gas pressure is 10 Pa   B) Applied voltage is 1.0 kV and Gas pressure is 30 Pa.

films prepared on the anode as shown Figure 6. It was remarkable that roughness was as deep as 270 nm when the film was prepared at 1.0 kV and 30 Pa on the anode. Furthermore, this film showed the ultra water repellent behavior, where the contact angle was as high as $162°$. Thus, a clear relationship was seen between the surface roughness and the water repellent behavior as indicated in Figure 7 (11-13)

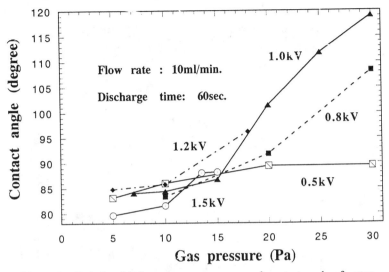

Figure 4.   Relationship between gas pressure and contact angle of water for various applied voltages.

Conclusion
        Plasma polymerized films of vinylidene fluoride were successfully prepared by the D. C. glow discharge system under various conditions.   Each PPVF film possessed almost the same chemical structures as judged from the IR spectra and the The water repellence of the films indicated a clear relationship with the surface roughness.   The maximum value of the water contact angle was $162°$, which is accepted as ultra water repellent.

References
1. Suwa, T.; Jikei, M. ; Kakimoto, M.; Imai, Y.; *Jpn. J. Appl. Phys.*, **1995**, *34*, 6503.
2. Yase, K.; Horiuchi, S.; Kyotani, M.; Yamamoto, K.; Yaguchi, A.; Futaesaku, Y.; Suwa, T.; Kakimoto, M.; Imai, Y.; *Jpn. J. Appl. Phys.*, **1996**, *35*, L567.
3. Suwa, T.; Jikei, M.; Kakimoto, M.; Imai, Y.; Tanaka, A.; Yoneda, K.; *Thin Solid Films*, **1996**, *273*, 258.
4. Sato, D.; Suwa, T.; Kakimoto, M.; Imai, Y.; *J. Photopolym. Sci. Tech..*, **1997**, *10*, 149 .
5. Hozumi, A.; Kakinoki, N.; Asai, Y.; Takai, O.; *J. Mater., Sci., Lett.*, **1996**, 15, 675.

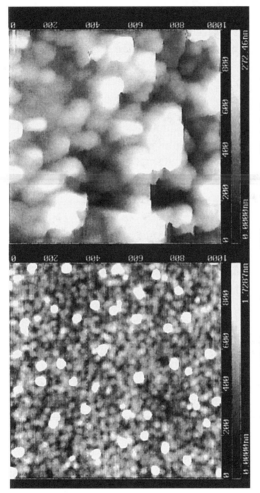

Figure 6. AFM images (scan area 1X1 μm, depth profile is indicated below the images) of PPVF films prepared on the anode. A) Applied voltage is 1.0 kV and Gas pressure is 10 Pa B) Applied voltage is 1.0 kV and Gas pressure is 30 Pa.

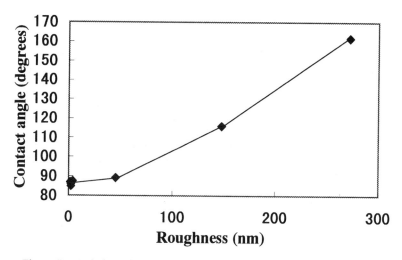

Figure 7.   Relationship between the roughness of the film surface and the water contact angle.

6. Hozumi, A.; Kondo, T.; Kajita, I.; Sekiguchi, H.; Sugimoto, N.; Takai, O.; *Jpn. J. Appl. Phys.*, **1997**, *36*, 4959.
7. Hozumi, A.; Sekiguchi, H.; Kakinoki, N.; Takai, O.; *J. Mater., Sci.*, **1997**, *32*, 4253.
8. Takai, O.; Hozumi, A.; Sugimoto, N.; *J. Non-crystal. Solids*, **1997**, *218*, 280.
9. Hozumi, A.; Takai, O.; *Thin Solid Films*, **1997**, *303*, 222.
10. Okada, Y.; *Thin Solid Films*, **1980**, *74*, 69.
11. Dettre, R. H.; R. E. Johnson, Jr., R. E.; *Adv. Chem. Ser.*, **1964**, *112*, 136.
12. Shibuichi, S.; Onda, T.; Satoh, N.; Tsujii, K.; *J. Phys. Chem.*, **1996**, *100*, 19512
13. Onda,T.; Shibuichi, S.; Satoh, N.; Tsujii, K.; *Langmuir*, **1999**, *12*, 2125

Chapter 13

Surface Interactions of CF_2 Radicals during Deposition of Amorphous Fluorocarbon Films

Neil M. Mackie[1], Nathan E. Capps[2], Carmen I. Butoi[3], and Ellen R. Fisher[3]

[1]Mattson Technologies, 3550 West Warren Avenue, Fremont, CA 94538
[2]Advanced Energy, Inc., 1625 Sharp Point Drive, Fort Collins, CO 80525
[3]Department of Chemistry, Colorado State University, Fort Collins, CO 80523-1872

Fluorocarbon polymer film composition and CF_2 radical-surface interactions in two fluorocarbon plasmas, 100% CHF_3 and 50/50 C_2F_6/H_2, are compared. Our imaging of radicals interacting with surfaces (IRIS) technique was used to collect spatially-resolved laser-induced fluorescence (LIF) images of CF_2 radicals interacting during film growth on silicon substrates. Simulation of LIF cross-sectional data shows that 100% CHF_3 plasmas have high scattering coefficients for CF_2 radicals. These large scattering coefficients (1.65 ± 0.03) indicate CF_2 molecules are generated through plasma interactions with the substrate. In contrast, IRIS data for CF_2 from 50/50 C_2F_6/H_2 show lower scattering coefficients (0.84 ± 0.02). Films deposited during IRIS experiments were characterized using X-ray photoelectron spectroscopy (XPS) and Fourier transform infrared (FTIR) spectroscopy and were nearly identical in composition regardless of plasma system. In addition, using independent plasma reactors, fluorocarbon polymer film deposition rates were measured as a function of applied rf power for both systems and optical emission spectra (OES) were collected and compared. The role of CF_2 radicals in fluorocarbon film deposition is discussed in relation to previously proposed film formation models.

Introduction

Fluorocarbon plasmas (FCPs) have been extensively studied because of their dual ability to promote etching of a variety of substrates[1,2] and to deposit a wide range of amorphous fluorinated carbon (a-C:F) polymer films.[3,4] Polymeric fluorocarbon materials deposited from FCPs exhibit a range of advantageous properties. Namely, their good adhesion to many organic and inorganic substrates; low surface free energies and inertness; low dielectric constants (k - 1.8-2.2);[5] high biocompatibility; and structural compositions that can be changed over a broad range producing films that are specifically tailored for each application.

While an enormous body of work using fluorocarbon plasmas (FCPs) exists, details of the mechanisms for fluorocarbon polymer deposition remain unclear. Gas-phase density measurements have shown there are significant concentrations of neutral radicals and energetic ions in FCPs. These active species are important chemical precursors to etching and deposition processes.[6,7,8] One mechanism for the deposition of fluorocarbon films from FCPs comes from d'Agostino and coworkers who propose an activated growth model (AGM).[5] AGM implies that film growth occurs through interaction of CF_x radicals with an "activated" surface. This activation occurs from charged particle bombardment creating reactive sites on the surface. This mechanism was postulated after d'Agostino and coworkers observed that fluorocarbon deposition rates were proportional to CF_2 radical and electron density in FCPs.[9] It has also been determined that the deposition of fluorocarbon films from FCPs displays a non-Arrhenius behavior and that there is a negative apparent activation energy for the deposition process.[10] This suggests adsorption-desorption processes are important in film growth.

In addition to the AGM, another factor that governs the etching and deposition capacity of FCPs is the ratio of CF_x radicals to F atoms. There are a number of operating conditions that modulate this ratio: fluorocarbon starting materials with high F/C ratios produce less CF_x radicals than F atoms;[11] and the addition of hydrogen to the fluorocarbon feed drives polymerization reactions forward by directly reacting with F to form stable HF and to prevent CF_x recombination reactions.[5] Surface interactions of H atoms in FCPs are not completely understood, although surface abstraction of F by H is thermodynamically possible.[12] Thus, the surface interactions of CF_x radicals as well as the role of fluorine and hydrogen during fluorocarbon polymer film deposition are critical for identification of species that contribute to film growth and for elucidation of plasma mechanisms.

There are few studies that have directly investigated the surface interactions of CF_x radicals during polymer deposition. Some studies have, however, correlated gas-phase species density with film growth rates. Kitamura and coworkers used laser-induced fluorescence (LIF) to detect CF_2 in C_2F_6 plasmas.[13] They observed a third order dependence of the CF_2 normalized LIF intensity to the polymer deposition rate with increasing applied rf power. Spatially resolved LIF measurements of CF_2 intensity radially across the deposition electrode also showed a strong correlation with CF_2 intensity and polymer film thickness. Under low growth rate conditions (100% CF_4 plasmas), Booth and coworkers have measured steady-state concentration profiles for CF and CF_2 radicals using different substrate materials.[14,15] Their spatially-resolved radial measurements suggest there is small production of CF_2 radicals on Al and SiO_2 substrates, and much higher surface production on Si substrates. They also find that energetic ion bombardment is a necessary component of the observed radical production.[15,16]

Our imaging of radicals interacting with surfaces (IRIS) technique directly measures the steady-state surface reactivity of a gas-phase species during plasma processing. IRIS combines molecular beam and plasma technologies with laser-induced fluorescence (LIF) to provide spatially resolved 2D images of radical species involved in film formation or etching.[17] To date, the IRIS method has been used to investigate seven different radical species[18,19,20] in several different plasma systems.[21,22] Most recently, we have focused on the surface reactivity of CF_x radicals in FCPs.[23,24,25] Here, in addition to surface reactivity data for CF_2 radicals in CHF_3 and 50/50 C_2F_6/H_2 plasmas, we present optical emission spectroscopy (OES) of the FCPs. Moreover, we correlate these gas-phase measurements with X-ray photoelectron spectroscopy (XPS) and Fourier transform infrared (FTIR) spectroscopy analyses of fluorocarbon films deposited during IRIS experiments. Thus, this fairly complete description of the plasma systems allows us the opportunity to offer insight into the plasma deposition mechanism.

Experimental Methods

The IRIS method has been described in detail previously.[17] In a typical IRIS experiment, feed gases enter the glass tube source region, rf power is applied, and a plasma is produced. Expansion of the plasma into a differentially pumped vacuum chamber, and ultimately into a high vacuum region, generates an effusive molecular beam consisting of virtually all species present in the plasma, including the species of interest. A tunable laser beam intersects the molecular beam downstream from the plasma source at a 45° angle and excites the radical of choice. Spatially resolved LIF signals are collected by an electronically gated, intensified charge coupled device (ICCD) located perpendicular to both the molecular beam and the laser beam, directly above the interaction region. A substrate is rotated into the path of the molecular beam and LIF signals are again collected. Differences between the spatial distributions with the surface in and out of the molecular beam are used to measure a radical's surface reactivity.

We can interpret the spatially-resolved LIF data with our quantitative model of the experiment,[18] which reproduces the scattering data in one dimension.[19] Briefly, the simulation is based on the known geometry of the experiment and calculates the spatial distribution of the radical number density in the molecular beam at the interaction region. Our model also calculates the radical number density along the laser beam for molecules scattering from the substrate surface assuming adsorption-desorption scattering. The calculated curve for this type of scatter assumes all of the incident radicals leave the substrate surface with a cosine distribution about the surface normal. To determine the surface reactivity of a specific molecule, the fraction of radicals scattering from the surface, S, is adjusted to best fit the experimental data. We define surface reactivity, R, as 1-S. R can be considered as an effective surface reaction probability for a given species.[26] Here, we will discuss the surface interactions of CF_2 primarily from the standpoint of scatter, S.

The source of the molecular beam is a continuous wave (CW) plasma consisting of 100% CHF_3 (Air Products, 99.95%), or a 50/50 mixture of C_2F_6 (Air Products 99.995%) and H_2 (General Air 99.99%). Total pressure in the source is 60-80 mTorr CHF_3 and 25-30 mTorr for C_2F_6/H_2 as measured by a capacitance manometer. The plasma is produced by the inductive coupling of 13.56 MHz rf power (50 or 100 W) tuned by an rf matching network. The molecular beam was collimated by two rectangular slits, 1.03 and 1.25 mm wide, mounted on a liquid nitrogen cold shield maintained at -200 °C.[20]

Tunable laser light was produced by frequency doubling the output of an excimer-pumped (XeCl, 100 mJ, 100 Hz) dye laser system with Coumarin 47 (CF_2, 229-240 nm, 1.3 mJ/pulse). Substrates are p-type silicon (100) wafers with ~50 Å of native oxide placed 2.0 - 4.0 mm from the laser beam. For all CF_2 reactivity experiments shown here, the laser was tuned to 234.323 nm, corresponding to the $(0,11,0)$ - $(0,0,0)$ vibronic band of the A^1B_1 - X^1A_1 transition. LIF from the A^1B_1 $(0,11,0)$ state of CF_2 was collected and focused directly onto the ICCD camera by two fused silica lenses as described previously.[23,24] All IRIS measurements are made in the saturated regime of laser power such that no corrections for fluctuations in laser power were made. Since the IRIS method is based on normalizing the signal from scattered radicals to the signal from radicals in the incident molecular beam, our measurements are not adversely affected by small fluctuations in laser power.

LIF signals were collected for 5 accumulations of 10-30 s exposure each for CF_2 and multiple sets of data were taken for each experiment. Background images were taken with the laser tuned to an off-resonance frequency and were subtracted from the data image. A 1-D representation of the image was made by averaging 20 columns of pixels (7.74 mm) containing the LIF signal and plotting signal intensity as a function of distance along the laser beam path. The ICCD camera had a 0.3 μs gate width for CF_2 molecule (radiative lifetime 61 ± 3 ns[27]).

Reflection IR spectra of deposited films were obtained with a Nicolet Magna 760 FTIR at 2 cm^{-1} resolution. XPS analyses were performed on a Surface Science Instruments S-probe spectrometer at the NESAC/BIO center at the University of Washington, which has a monochromatic Al Kα X-ray source ($h\nu$ = 1486.6 eV), hemispherical analyzer, and resistive strip multichannel detector. Details of the apparatus configuration and correction of the binding energy (BE) scales for these measurements are given elsewhere.[24]

Optical emission spectra (OES) from 240 - 713 nm were obtained for the FCPs using an inductively coupled plasma reactor nearly identical to the one used as the molecular beam source in IRIS experiments.[28] Emitted light was collected through a quartz window situated 10 mm downstream of the coil region, resulting in coaxial sampling of the plasma emission. Emission was imaged onto the 10 μm entrance slit of an Ocean Optics S2000 spectrograph using an optical fiber. The spectrograph was equipped with a 1800 grooves/mm holographic grating and a 2048-element linear CCD-array detector. Deposition rates were measured using a Tencor alpha step 100 profilometer. Details on deposition rates for the C_2F_6/H_2 system are provided in Ref. 28.

Results

LIF Excitation Spectra of CF_2. LIF measurements of CF_2 radicals were made in the IRIS apparatus with no substrate in the path of the molecular beam.

LIF provides a direct spectroscopic tool for identification and independent study of a specific ground-state radical in a molecular beam populated with many species. Figure 1 shows excitation spectra of CF_2 from 50/50 C_2F_6/H_2 and 100% CHF_3 plasmas (50 W) taken by stepping the laser from 227 to 240 nm in 0.025 nm increments (1s/step). Comparison to literature spectra verifies that the fluorescing species is indeed CF_2,[27] and demonstrates there are no contributions from any other molecular beam species in this wavelength range. Therefore, we are assured that the LIF signals we collect for subsequent experiments are solely LIF from CF_2 radicals.

Optical Emission Spectroscopy. OES is also a useful spectroscopic tool that allows for the identification of excited state species in a plasma. Here, we collect emission spectra for our two plasma systems using the independent plasma reactors described previously.[28] Plasma emission in the range 240 to 713 nm was analyzed from 100% CHF_3 and 50/50 C_2F_6/H_2 plasmas (50 W), Figure 2. The spectrum of the 50/50 C_2F_6/H_2 plasma is dominated by CF_2 emission from the bending mode of the A^1B_1 - X^1A_1 system, characterized by vibronic transitions $A(0,v_2, 0)$ - $X(v_1, v_2, 0)$ from 249 to 300 nm, Figure 2a.[5] The small peak at 387 nm is assigned to N_2 emission, which is a residual component of our plasma reactor. In addition, there are peaks attributed to H_α and H_β atomic emission at 656.3 nm and 486.1 nm respectively.[29] The 100% CHF_3 emission spectrum is similar, exhibiting a strong CF_2 emission, Figure 2b. One major difference, however, is that the H_α emission is significantly less intense than observed for the 50/50 C_2F_6/H_2 plasma

Surface Reactivity of CF_2. Figure 3 shows a series of ICCD images of CF_2 using a 100% CHF_3 plasma molecular beam and a Si substrate. The LIF signal from CF_2 molecules in the molecular beam is shown in Figure 3a. Figure 3b is the LIF signal acquired with a Si substrate rotated into the path of the molecular beam. Here, both the incident molecular beam and scattered CF_2 molecules are imaged. Figure 3c is the difference between Figures 3b and 3a and shows only CF_2 molecules scattered off the surface of the substrate.

Figure 4 shows the data of Figures 3a and 3c converted to a 1-D graphical representation for radicals both in the incident molecular beam and scattered, or desorbed, CF_2 molecules. The broad spatial distribution and the shift of the scattered signal peak maximum away from the molecular beam peak maximum indicate the CF_2 radicals scatter with a cosine angular distribution. Also shown in Figure 4 are simulated curves (dashed lines) for the incident beam and for scattered molecules assuming an adsorption-desorption mechanism, with $S = 1.65 \pm 0.05$. A scattering value greater than unity indicates that surface production of CF_2 (g) contributes to the observed scattering signal. Scattering values for CF_2 on other substrates, SiO_2, Si_3N_4, photoresist, and 304 stainless steel, are similar to that determined for silicon.[24]

Figure 5 shows cross sectional LIF data for CF_2 using a 50:50 C_2F_6/H_2 plasma molecular beam processing a silcon substrate. Here, the relative intensity of the scattered signal has decreased significantly from the 100% CHF_3 data. Simulating the 50:50 C_2F_6/H_2 experiment as above yields $S = 0.86 \pm 0.09$. This is a considerably lower scatter than was observed above with a 100% CHF_3 plasma

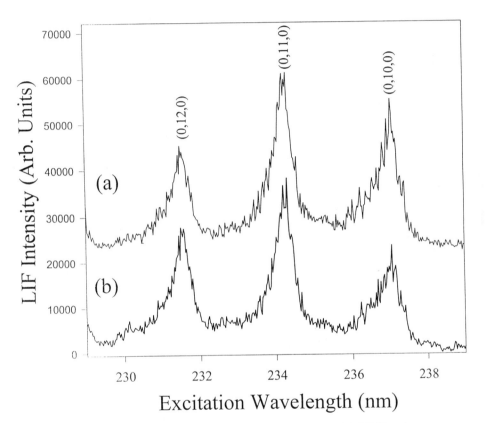

Figure 1. Experimental fluorescence excitation spectra of CF_2 in (a) 50/50 C_2F_6/H_2 and (b)100% CHF_3 plasma molecular beam from 227 to 239 nm. The transition used for all reactivity experiments was the (0,11,0) vibronic band of the $A^1B_1 - X^1A_1$ transition.

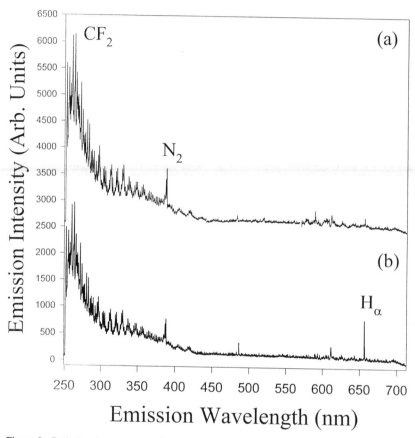

Figure 2. Optical emission spectra of (a) 50/50 C_2F_6/H_2 and (b)100% CHF_3 plasma from 249 to 713 nm. Spectra collected coaxially from an independent inductively coupled rf plasma reactor.

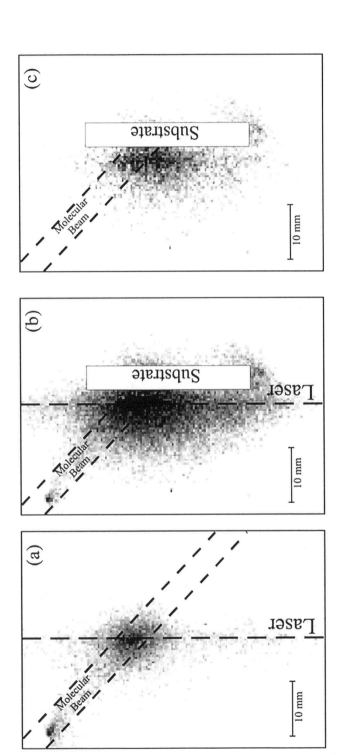

Figure 3. Spatially resolved two-dimensional ICCD images of the LIF signal for the CF_2 (0,11,0) state (a) in the 100% CHF_3 molecular beam (no substrate) and (b) with a 300 K Si substrate rotated into the path of the molecular beam at a laser-surface distance of 3 mm. (c) Difference between the images shown in (a) and (b), corresponding to CF_2 molecules scattering from the surface. LIF signals with the highest intensity appear as the darkest regions in the images. Dashed lines indicate the location of the molecular beam and the laser beam.

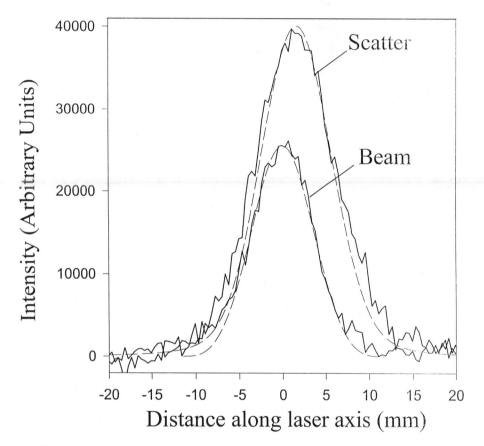

Figure 4. Cross-sectional data for the LIF of CF_2 in the incident molecular beam and scattered from the Si (100) substrate using a 100% CHF_3 plasma molecular beam. The laser-surface distance is 3.0 mm. Dashed lines represent the simulated curves from the geometric model, assuming $S = 1.65$ and adsorption-desorption scattering.

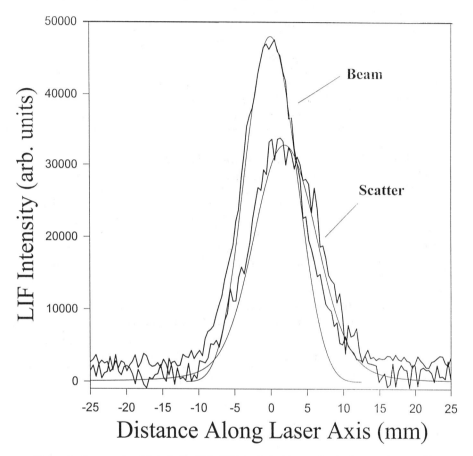

Figure 5. Cross-sectional data for the LIF of CF_2 in the incident molecular beam and scattered from the Si (100) substrate using a 50/50 C_2F_6/H_2 plasma molecular beam. The laser-surface distance is 3.0 mm. Dashed lines represent the simulated curves from the geometric model, assuming $S = 0.84$ and adsorption-desorption scattering.

molecular beam and this polymeric substrate. This result indicates that in this FCP system, CF_2 is removed from the gas phase during plasma polymerization. Values for CF_2 scatter from SiO_2, and polyimide surfaces using a 50/50 C_2F_6/H_2 plasma were similar to the value determined for Si.[25]

Deposition rates of fluorocarbon films. Due to low molecular beam species flux and corresponding low deposition rates in our IRIS experiment, deposition rates were determined for films deposited from 50/50 C_2F_6/H_2 and 100% CHF_3 in an inductively coupled plasma reactor nearly identical to the source on IRIS.[28] In general, deposition rates for C_2F_6/H_2 are five times faster than for films deposited from CHF_3 under identical conditions, Fig. 6. Deposition rates from C_2F_6/H_2 plasmas increase rapidly from 10 to 30 W to a value of 770 Å/min. The deposition rate goes through a maximum at 100 W where it reaches 1200 Å/min and decreases at higher powers to ~800 Å/min. In contrast, the deposition rate from CHF_3 plasmas increases initially, then remains constant at a rate of 100 Å/min from 25-150 W applied power. The deposition rate then increases to a value of 400 Å/min at the highest applied power used, 200 W.

Characterization of fluorocarbon films. To correlate our radical surface reactivities with the composition of the deposited material, IRIS substrates were removed from the instrument after reactivity measurements were completed and surface analysis was performed. Figure 7 shows C_{1s} spectra for films deposited on silicon substrates during IRIS experiments using 50/50 C_2F_6/H_2 and CHF_3 plasmas. Analysis shows the experimental curves can be fit using a series of peaks assigned to CF_3, (BE = 294 eV), CF_2 (BE = 292.1 eV), CF (BE = 289.5 eV), C- CF (BE = 287.3), and CH (285 eV).[5] Elemental composition and C1s contributions for these films are listed in Table I.

Table I. XPS C1s and elemental composition for films deposited on Silicon from 100% CHF_3 and 50/50 C_2F_6/H_2 plasmas

| Plasma | C1s Contribution % | | | | | | Elemental Composition % | | | | |
	CH	C-CF	CF	CF_2	CF_3	F/C	F	C	O	N	Si
100% CHF_3	17	29	27	20	7	0.8	41.4	50.7	5.4	2.6	-
50/50 C_2F_6/H_2	13	29	27	19	12	1.0	48.2	48.4	3.1	-	0.3

Figure 8 shows FTIR spectra of amorphous fluorocarbon polymers deposited on Si substrates from 100% CHF_3 and 50/50 C_2F_6/H_2 plasmas. Films deposited from 50/50 C_2F_6/H_2 plasmas during IRIS experiments were analyzed directly. To obtain an acceptable FTIR absorbance, films were deposited from 100% CHF_3 in an

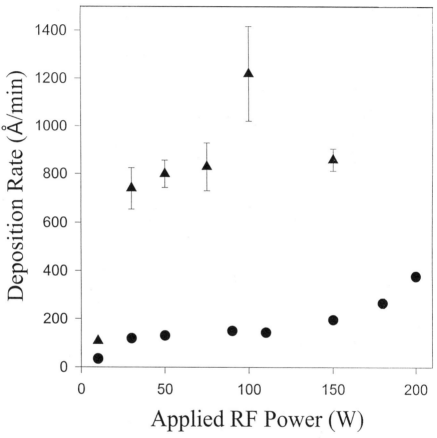

Figure 6. Deposition rates for films deposited for 10-15 minutes from 50/50 C_2F_6/H_2 (closed triangles) and 100% CHF_3 (closed circles) plasmas. All films were deposited in an independent inductively coupled rf plasma reactor.

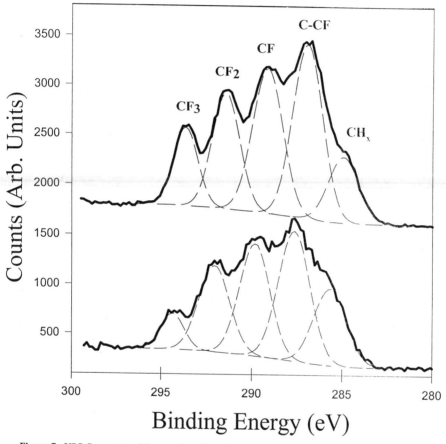

Figure 7. XPS C_{1s} spectra of fluorocarbon films deposited during IRIS experiment using (a) 50/50 C_2F_6/H_2 and (b) 100% CHF_3 plasma molecular beam. Spectra were taken at a photoelectron take-off angle of 55° from the surface normal.

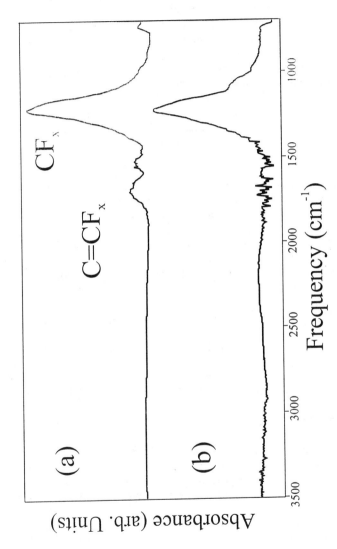

Figure 8. FTIR transmission spectra of fluorocarbon films deposited from a (a)100% CHF_3 50/50 plasma from an independent inductively coupled rf plasma reactor, and (b) C_2F_6/H_2 plasma during an IRIS experiment.

independent inductively coupled plasma reactor described in the previous section. Conditions for these depositions were 50 W applied rf power and 100mTorr pressure in the reactor. The film deposited from 100% CHF_3 shows two major absorption bands, one at 1100 - 1400 cm^{-1} and another, smaller peak at ~1700 cm^{-1}, Fig. 8a. These peaks are attributed to CF_x (x = 1-3) and unsaturated stretching modes, respectively.[30] Aydil and coworkers have observed this type of unsaturation in films deposited from CF_4/H_2 plasmas under deposition conditions, i.e. >25% H_2. Under etching conditions, the degree of unsaturation decreases significantly.[31] Further analysis of the spectra reveal a A(C=C)/A(C-F) ratio of 0.20 ± 0.01 for the 50 W data. This is consistent with the results of Aydil and coworkers for conditions where the fluorocarbon layer thickness increases with time. Amorphous fluorocarbon films are also deposited from 50/50 C_2F_6/H_2 and are similar to films deposited from CHF_3 plasmas in that they contain an amorphous CF_x band at 1100 - 1400 cm^{-1}, however, the degree of unsaturation in these fluorocarbon polymer films is much lower.

Discussion

Although FCPs have been studied extensively, past studies have rarely discussed both gas-phase characterization and substrate processing effects. Indeed, most researchers concentrate on only one aspect of these complicated systems. Yet, determining how gas-phase radicals interact with a substrate is critical to understanding the chemistry of plasma processing. This is especially important in FCPs as the balance between film growth and etching depends on the interactions of active plasma species with the substrate being processed. Insight into the reactions of radicals with a surface *during* plasma processing under different plasma conditions can be obtained with IRIS measurements. IRIS allows the study of one specific molecule interacting with a specific surface while it is being bombarded by the full range of plasma species.

Comparison of the two fluorocarbon systems studied here illuminates the similarities and differences in the chemical and physical process that occur during plasma polymerization of fluorocarbon materials. First, amorphous fluorocarbon polymer films are deposited from both 100% CHF_3 and 50/50 C_2F_6/H_2 plasmas. Our XPS and FTIR analyses reveal the chemical composition of these materials are very similar and indicate a high degree of crosslinking which is likely the result of ion bombardment.[28,32] Deposition rates in the two systems, however, are significantly different. Indeed, the deposition rate for the 50/50 C_2F_6/H_2 system is at least a factor of four higher than in the CHF_3 system. Second, from our OES data, we find that both precursors decompose to generate CF_2 and H atoms, although in very different amounts, with the higher CF_2 radical and H atom content occurring in the C_2F_6/H_2 system.

Most significantly, the surface interactions of CF_2 radicals during polymer deposition measured with IRIS are very different in the two systems. Our IRIS data clearly indicate we have consumption of CF_2 (S = 0.86 ± 0.09) on Si substrates using 50/50 C_2F_6/H_2 plasmas. Consumption of radicals in IRIS experiments during plasma deposition is generally thought to indicate the radical species is contributing to film growth.[17] In contrast, we have a high degree of scatter for CF_2 radicals (S = 1.65 ± 0.05) on Si substrates using 100% CHF_3 plasmas. This demonstrates that CF_2 is generated at the surface during film deposition in this system. We have previously discussed several possible surface reactions that could account for the observed production of CF_2 radicals in FCPs.[23,24] Briefly, these include dissociative adsorption of molecular beam species, such as CF_3 (producing CF_2 and an adsorbed F atom); surface F atom abstraction by CF radicals; and ion assisted sputtering of the deposited fluorocarbon polymer. Based on results from experiments employing a grounded mesh screen to remove ions from our molecular beam, we believe ion bombardment is a significant source of CF_2 in the CHF_3 system.[24]

The observations that high CF_x content and ion bombardment affect CF_2 surface reactivities are consistent with d'Agostino and coworkers' proposed activated growth model (AGM) for the formation of polymeric films in FCPs.[5] One of the key points of the AGM is that a relatively high density of fast electrons or energetic ions is necessary for the growth process. These high energy species are responsible for the formation of activated polymer sites on the growing film surface. Indeed, d'Agostino and coworkers found that deposition kinetics depend on both CF_x (x = 1, 2) concentration and on the flux and/or energy of positive ions.[5] Moreover, they observed that when the energy of ions is roughly equivalent to that of a floating or a biased substrate (i.e. with voltages < 50-70 eV), deposition rates increased with both fluxes of radicals and ions (or simply with ion energy).[33] With more energetic ions, however, ion-assisted etching starts to compete with deposition; thus, the deposition rate goes through a maximum when plotted against the bias voltage.[34]

As noted above, the H atom content is significantly greater in the C_2F_6/H_2 plasma, even though the C/F/H mole ratios (1:3:1) in the two FCP systems are identical. It is well known that the deposition rate for fluorocarbon polymers from saturated perfluorinated precursors is directly related to the amount of H_2 added to the system, with a maximum at ~50% H_2 addition. Moreover, Aydil and coworkers have shown that fluorocarbon polymers exposed to a H_2 plasma are surface modified via replacement of F atoms in the polymer by H atoms.[32] This replacement occurs through abstraction of surface fluorine by hydrogen and subsequent reaction between gas-phase H atoms and the newly created active surface site. In the C_2F_6/H_2 system, not only do we have H atom flux to the surface, we also have CF_x flux to the surface. Thus, it is conceivable that the active surface sites created by F abstraction react with both H atoms and CF_x radicals, resulting in a relatively rapid deposition rate. In the CHF_3 system, the H atom flux to the surface is lower, as is the CF_x flux, therefore we observe a much slower deposition rate.

FCPs are unique in that the relative importance of active species (CF_x radicals and energetic charged particles) can be continuously varied by controlling the plasma parameters and feed gas content. This means that the etching or polymerizing capacity of FCPs can be tailored to suit a particular application. This is exemplified by the two systems we study here. In the 50/50 C_2F_6/H_2 system, the deposition rate is at a maximum,[28] implying the active species contribute primarily to deposition and etching is minimized. This suggests that ion bombardment of the surface is only serving to create active sites for film growth, and is not ablating a significant amount of the deposited material. Consequently, in our IRIS experiments, we observe net consumption of CF_2, rather than generation. In the CHF_3 system, the active species are also contributing to deposition, but at a much slower rate. This suggests that both etching and deposition are occurring at nearly equal rates. Thus, ion bombardment of the surface serves primarily to ablate the depositing fluorocarbon material rather than to create active sites. This is also supported by our IRIS results for CF_2 in a primarily etching system, 100% C_2F_6 plasmas. In this system, CF_2 was generated at the surface of a Si substrate ($S = 1.44$ ± 0.03) and no net deposition was measured.

Conclusions

Fluorocarbon deposition mechanisms have been investigated using our unique IRIS technique in combination with extensive surface analysis of deposited films and optical emission spectroscopy of the C_2F_6/H_2 and CHF_3 plasmas. IRIS measurements for CF_2 radicals in the two plasma systems demonstrate two very different types of surface interactions. In the C_2F_6/H_2 plasma, CF_2 radicals are consumed during rapid deposition of a thick fluorocarbon polymer film. In contrast, in the CHF_3 system, CF_2 radicals are generated at the surface during much slower film deposition. These differences in surface reactivity and deposition rates are explained by consideration of the relative fluxes of active species in both systems and by the role of H atoms in the passivation of the fluorocarbon film. Additional clarification of the role of charged species as well as IRIS measurements for other CF_x radicals are clearly needed to further understand the chemistry of FCPs.

Acknowledgments

We thank Prof. David G. Castner of the University of Washington for performing XPS analyses on our fluorocarbon films. This work is supported by the National Science Foundation (CHE-951157).

References

1. Miyata, K.; Hori. M.; Goto, T. *J. Vac. Sci. Technol. A* **1997**, *15*, 568.

2. Inayoshim M.; Iti, M.; Hori, M.; Goto, T.; Hiramatsu, M. *J. Vac. Sci. Technol. A* **1998**, *16*, 233.

3. Kadono, M.; Inoue, T.; Miyanaga, A.; Yamazake, S. *Appl. Phys. Lett.* **1992**, *61*, 772.

4. d'Agostino, R.; Lamendola, R.; Favia, P.; Giques, A. *J. Vac. Sci. Technol. A* **1994**, *12*, 308; Sah, R.; Dishler, E. B.; Bubenzer, A.; Koidl, P. *Appl. Phys. Lett.* **1985**, *46*, 739.

5. d'Agostino, R.; Cramarossa, F.; Fracassi, F.; Illuzzi F. in *Plasma Deposition, Treatment and Etching of Polymers*, R. d'Agostino, Ed., Academic Press: San Diego, CA, **1990**, pp. 95.

6.   Suzuki, C.; Kadota, K. *Appl. Phyd. Lett.* **1995**, *67*, 2569.

7.   O'Neil, J. A.; Singh, J. *J. Appl. Phys.* **1994**, *76*, 5967.

8.   Takahashi, K.; Hori, M.; Goto, T. *J. Vac. Sci. Technol. A* **1996**, *14*, 2004.

9.   d'Agostino, R.; Cramarossa, F.; Illuzzi, F. *J. Appl. Phys.* **1987**, *61*, 2754.

10.   d'Agostino, R.; Cramarossa, F.; Fracassi, F.; DeSimoni, E.; Sabbarini, L.; Zambonin, P. G.; Caporicco, G. *Thin Solid Films* **1986**, *143*, 163.

11.   Hikosaka, Y.; Toyoda, H.; Sugai, H. *Jpn. J. Appl. Phys.* **1993**, *32*, L690; O'Keefe, M. J.; Rigsbee, J. M.; *J. Appl. Polym. Sci.* **1994**, *53*, 1631.

12.   Takahashi, K.; Hori, M.; Maruyama, K.; Kishimoto, S.; Goto, T. *Jpn. J. Appl. Phys.* **1993**, *32*, L694.

13.   Kitamura, M. Akiya, H, Urisu, T. *J. Vac. Sci. Technol. B.* **1989**, *7*, 14.

14.   Booth, J. P.; Cunge, G.; Chabert, P.; Schwarzenbach, W. In *Frontiers in Low Temperature Plasma Diagnostics II*, Bad Honnef, Germany, **1997**, 147.

15.   Tserepi, A. D.; Derouard, J.; Booth, J. P.; Sadeghi, N. *J. Appl. Phys.* **1997**, *81*, 2124.

16.   Cunge, G. Ph.D. Thesis, **1997**, Laboratoire de Spectrometrie Physique, Université Joseph Fourier, BP 87, 38402 St Martin d'Hères Cedex; J. P. Booth, private communication.

17.   McCurdy, P. R.; Bogart, K. H. A.; Dalleska, N. F.; Fisher, E. R. *Rev. Sci. Instrum.* **1997**, *68*, 1684.

18.   Bogart, K. H. A.; Cushing, J. P.; Fisher, E. R. *Chem. Phys. Lett.* **1997**, 267, 377.

19.   Bogart, K. H. A.; Cushing, J. P.; Fisher, E. R. *J. Phys. Chem.* **1997**, *101*, 10016.

20.   McCurdy, P. R.; Venturo, V. A.; Fisher, E. R. *Chem. Phys. Lett.* **1997**, *274*, 120.

21.   Buss, R. J.; Ho, P. *IEEE Trans. Plasma Sci.* **1996**, *24*, 79.

22.   Buss, R. J.; Ho, P.; Weber, M. E. *Plasma Chem. Plasma Process.* **1993**, *13*, 61.

23.   Mackie, N. M.; Venturo, V. A. Fisher, E. R. *J. Phys Chem* **1998**, *101*, 9425.

24.   Capps, N. E.; Mackie, N. M. Fisher, E. R. *J. Appl. Phys.* **1998**, *84*, 4736.

25.   Butoi, C. I., Mackie, N. M.; Williams, K. L.; Capps, N. E.; Fisher, E. R. *J. Vac. Sci. Technol. A,* submitted for publication.

26.   Perrin, J.; Shiratani, M.; Kae-Nune, P.; Videlot, H.; Jolly, J.; Guillon, J. *J. Vac. Sci. Technol. A* **1998**, *16*, 278.

27.   King, D. S.; Schenck, P. K.; Stephenson, J. C. *J. Mol. Spectrosc.* **1979**, *78*, 1.

28.   Mackie, N. M.; Dalleska, N. F.; Castner, D. G.; Fisher, E. R. *Chem. Mater.* **1997**, *9*, 349.

29.   Barshilia, H. C.; Mehta, B. R.; Vankar, V. D. *J. Mater. Res.* **1996**, *11*, 2852.

30.   Colthup, N. B.; Daly, L. H.; Wiberley, S. E. *Introduction to Infrared and Raman Spectroscopy*, 3rd Ed.; Academic Press: New York, 1990.

31.   Marra D. C.; Aydil, E. S. *J. Vac. Sci. Technol. A* **1997**, *15*, 2508.

32.   Yasuda, H. *Plasma Polymerization*; Academic Press: Orlando, FL, 1985.

33.   d'Agostino, R., personal communication (1998).

Chapter 14

Chemical and Contact Mechanical Characterization of Thin Plasma-Deposited Hexafluoropropylene Films

Reto Luginbühl[1], Michael D. Garrison[1,2], René M. Overney[2], Lothar Weiss[3], Holger Schieferdecker[3], Sabine Hild[3], and Buddy D. Ratner[1,2]

[1]Engineered Biomaterials and [2]Chemical Engineering, University of Washington, Seattle, WA 98195–1750
[3]Abteilung für Experimentelle Physik, Universität Ulm, D–89069 Ulm, Germany

Hexafluoropropylene (HFP) was deposited in a radio frequency plasma enhanced chemical vapor deposition (RF-PECVD) process as a function of applied reactor energy. The resulting polymer films were investigated regarding chemical composition with the electron spectroscopy for chemical analysis (ESCA). Photolithographic masking of silicon with subsequent deposition of HFP plasma polymer allowed for creating of micropatterned plasma films. The films were inspected with the scanning force microscope (SFM) regarding film thickness and surface roughness. In addition, the adhesion forces between SFM tip and the patterned substrates were probed via Pulsed Force Microscopy (PFM). The chemical composition and mechanical properties were found to be dependent on the plasma polymerization conditions. These properties were compared and related to the measured adhesion forces.

A central hypothesis in modern biomaterial research proposes that precision engineering of biomaterial surfaces, utilizing specific recognition strategies, will trigger intended biological responses. To appropriately test such hypotheses, precise arrays of biorecognition motifs are required. Ideally, an engineered biomaterial will have mechanical and materials properties coupled to its intended in-vivo function, while the surface layer will be "activated" to direct the cellular response. This strategy of specific surface modification takes advantage of existing materials, and uses only the surface chemistry and architecture of the interfacial region.

One approach to engineered surface design involves coating a standard biomaterial surface with a thin, "bioactive" film. *Chemical vapor deposition* (CVD) is a well-established class of techniques used for such thin film modification. In particular, the method of *radio-frequency plasma enhanced chemical vapor deposition* (RF-PECVD), also known as *radio-frequency glow discharge* (RFGD), has proven useful for biomaterial applications. RF-PECVD has a number of advantages for engineering biomaterial surfaces including: control of surface chemistry and functional group presentation, production of pinhole-free, conformal surfaces, and solvent-free room temperature processing. Through careful selection of the monomer, films have been prepared by RF-PECVD that display a high degree of "bioactivity": (1) resisting bacterial and protein attachment (1), (2) resisting platelet activation (2,3), and (3) binding biomolecules with high affinity (4,5).

Hexafluoropropylene (C_3F_6, HFP) belongs to a class of fluorocarbon monomers that have been shown to bind biomolecules tightly. Recently the RF-PECVD polymerization mechanisms and processing parameters of HFP have been described in detail (6,7). Tailoring of the processing conditions allows for the generation of thin, smooth films (<100Å) with precise surface chemistry that promotes the tight binding of proteins. Through standard photolithographic means, it has been shown that RF-PECVD-HFP films can be patterned in precise geometries and architectures on the surface to produce regions of high binding within a lower affinity background (8-11). The phenomenon of tight protein binding and its modulation is of great importance in the development of an engineered surface to direct biological response. As a result RF-PECVD HFP films invite further study in terms of pattern development and quantitation of binding strength.

Measuring binding affinity of surfaces at the cellular and molecular size scale is a challenging task. Fortunately, advances in scanning force microscopy (SFM) have occurred to allow for the spatial mapping of contact mechanical surface behavior (12). Surface topography, friction coefficient, elastic and/or shear moduli are all accessible through the various operational modes of an SFM. In addition, the force of adhesion between the probe and the surface can also be spatially quantitated using a recently developed method called *pulsed force mode* (PFM) (13). By patterning the material of interest against a common surface (e.g. silicon), a set of materials can be compared relative to the standard.

In this work we discuss the relationship between the chemical composition and contact mechanical properties of RF-PECVD HFP films. By varying the reactor conditions, films of varying surface chemistry and crosslink density were prepared. Chemical composition and functional group presentation was measured *by electron spectroscopy for chemical analysis* (ESCA). Patterned films were prepared using photolithographic techniques. Surface morphology, pattern fidelity, and film thickness were measured by scanning probe microscopy. Quantitative spatial mapping of the surface adhesion to the scanning probe was measured using PFM. The probe-surface affinity was correlated with the degree of crosslinking in the surface film and the surface chemical composition.

Experimental Methodology

Cleaning and Photolitographic Patterning

Silicon wafers (100 orientation, Silicon Quest Intl., Santa Clara CA), were cut to 10 x 10 mm size and were cleaned in acetone and methanol, (analytical grade, from commercial sources) and dried in a stream of pre-purified nitrogen. The surface oxide layer was removed by dissolution in concentrated hydrofluoric acid (Fisher, 49% v/v). Finally the silicon surfaces were rinsed in ultrapure water (R=18MΩ, US Filter Corp.) and dried in nitrogen. As HF is highly dangerous, all HF cleaning was carried out in a well ventilated fumehood with extensive personal protection.

A photoresist pattern on the silicon surfaces was obtained by coating the chips with a positive resist (AZ1512, Shipley Inc.), irradiation through a photomask and subsequent development in a buffer solution (Shipley AZ351 developer). All substrates were stored in sample trays under air until mounted in the plasma reactor chamber. Following the plasma film deposition, the remaining photoresist was liftoff by re-developing (14). These fluoropolymer patterns were then rinsed in deionized water and stored sealed in sample trays under air until further use.

Radio Frequency Plasma Enhanced Chemical Vapor Deposition

In the RF-PECVD process energy is transferred by the radio electric field to electrons, which collide with the monomer molecules and the surface. Collisions with the molecules generate more electrons, free radicals and molecular ions and molecules in excited state. All fragments are very reactive and may react with each other or with the exposed surfaces. If the monomer is a "non etching" molecule, a polymer is formed on the surface. As the electrons do not reach the thermodynamic equilibrium with the monomer, the temperature of the gas molecules is close to ambient. This enables the use of thermolabile monomers without major degradation of the reactants and therefore quite often the monomer structure is retained in the plasma deposited polymer. An applied RF frequency of 13.56 MHz, results in a plasma polymer with a relative low crosslinking degree compared to plasma polymers formed with RF frequencies of 5 kHz (15).

The attractiveness of that method lies in its versatility and performance. Any material surface can be modified by plasma deposition. Quite often chemicals (monomers), which are described as non-polymerizable with standard methods proved to be polymerizable with RF-PECVD (16). The polymerization process and the deposition rate are highly dependent on the reaction parameters chosen. Upon proper choice of the reactor conditions the produced films are very smooth, continuous and pinhole free (9,17).

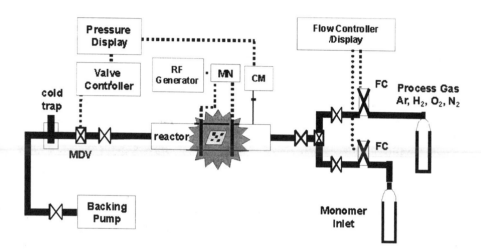

Figure1. Schematic of the Plasma Reactor for RF-PECVD.
In the low pressure reactor a plasma is induced of a gasous monomer by applying a radio field between two electrodes. Sample surfaces are placed directly in the plasma glow. MN = Matching Network, CM = Capacitance Manometer, FC = Flow Controller, MDV = Motor Driven Valve

For all experiments discussed herein, we used *hexafluoropropylene* (HFP, C_3F_6; PCR, Gainsville, FL) as monomer. RF-PECVD was carried out in a reactor tube equipped with symmetrical external, capacitively coupled electrodes (Figure 1). The separation distance between the electrodes was kept constant at 10 cm. All sample surfaces were placed in-between the electrodes directly in the plasma ("in the glow"). Base pressure in the evacuated reactor chamber was typically below 1 mTorr. A mass flow controller situated at the entrance of the reactor carefully controlled monomer flow and an automated gas valve at the reactor end maintained a constant monomer pressure. For all HFP plasma deposition experiments monomer flow and pressure conditions were kept constant at 2 sccm and 150 mTorr, respectively. The plasma was induced by a RF field of 13.56 MHz and maintained in a steady state condition by matching the impedance of the reactor to that of the RF generator via a matching network. In our experiments the applied reactor power was varied between 2 W and 100 W. Following to a 8 min plasma deposition the RF field was turned off and all samples were "annealed" in monomer flow for 5 minutes and then brought back to atmospheric pressure under nitrogen.

Electron Spectroscopy for Chemical Analysis (ESCA)

Analysis of the chemical composition and structure of the outermost 60-100 Å of a surface was performed by ESCA. This well established technique is especially useful for the analysis of plasma polymerized films in, which the monomer chemistry may be slightly retained. In particular, ESCA analysis of fluorocarbon films benefits from the large chemical shifts in the core level carbon (C 1s) window due to the strong electronegativity of the associated fluorine atoms.

ESCA analysis on unpatterned HFP plasma polymer films was performed on a Surface Science Instruments (SSI, Mountain View, CA) S-Probe ESCA instrument, utilizing an aluminum K_α1,2 monochromatized X-ray source and a hemispherical energy analyzer. This instrument permits the investigation of the outermost 20-100 Å of up to 1 mm^2 surface area of the sample at high (150 eV pass energy) or low (50 eV pass energy) resolution. Samples were mounted on a rotating carousel and a nickel grid positioned $\approx$ 3 mm above the sample surface. As the HFP films are highly insulating, surface charging of the non-conducting films was minimized with an electron flood gun set at 5eV and the nickel grid diffuser. Typical pressures in the analysis chamber during spectral acquisition were 10^{-9} Torr. Typically, a surface was interrogated with a low resolution scan, followed by a high resolution scan of the C 1s window, followed by another low resolution scan. This methodology allowed for the detection of film degradation due to the incident X-ray flux. Total analysis time per spot on a given sample was less than 45 minutes. At longer analysis times (t > 1.5hrs) degradation in the film fluorine to carbon ratio was observed between the first and final low resolution scans. SSI data analysis software was used to calculate the elemental compositions from the peak areas in low resolution spectra and peak fit the component peaks in the high resolution spectral envelope. For this study, all spectra were recorded at a take-off angle of 55°. Spectral binding energies were referenced to the CF_2 peak assigned to 292 eV (18).

Scanning Force Microscopy (SFM)

Since the invention of the *scanning force microscope* (SFM) in 1986 by Binnig et al (19) major technical advances have been made. SFM has become one of the most important instruments to study topographical features and contact mechanical properties at the micro- and nanometer scale. The SFM has unparalleled sensitivity to measure small, real space, variations in topography and allows for qualitative and quantitative characterization of surface forces at the nanometer scale.

All contact mode SFM experiments were carried out with an Explorer standalone system (TopoMetrix Inc.) equipped with the standard 100 μm x-y scanner and a 12 μm z-piezo. These investigations were carried out in a nitrogen flooded dry box (humidity typically below 4 % at 22 °C). Bar shaped rectangular silicon cantilevers were used with a resonance frequency between 160 and 175 kHz (low resonance non-contact tips, TopoMetrix Inc.). Tip-sample forces were minimized (≤ 10nN), and scan rates were typically 0.5 or 1 Hz. Normal forces were estimated from tip-sample force curve analysis. The manufacturer's nominal values for normal and lateral force constants were used. Raw images were flattened to correct for non-linear curvature of the sample stage piezo tube response.

Pulsed Force Mode

Elastic properties and adhesive forces of the surface can be determined by scanning force microscope (SFM) performing force vs. distance curves (FDC). In the repulsive part of the force curves, the slope of the cantilever response is determined by the elastic properties of a material. The adhesive properties are correlated to the pull off force measured when the tip is retracted from the surface (20). To obtain a two dimensional image of the adhesive force, force vs. distance curves may be taken and the pull-off force has to be estimated for each point (21). Based on this technique a new method has been developed, called *Pulsed Force Mode* (PFM) (13,22,23). In the Pulsed Force Mode force vs. distance curves are performed by a sinusoidal z-modulation of the piezo. The resulting deflection signal as a function of time is shown in Figure 2. Typical modulation frequencies are in the range from 500 Hz to 2 kHz using amplitudes between 10 nm and 1 micron. Thus, the normal contact time between tip and surface in PFM is about 10 μs. Therefore delicate samples can be measured in high resolution without or with reduced sample damage. The PFM works far below the resonant frequency of the cantilever, which allows an exact determination of the normal forces applied to the surface during the whole sine period. To perform the force curves in PFM an oscillator in the PFM-circuitry creates the sinusoidal modulation frequency, which is added to the z-scan piezo. The generated force signal is a repetitive measurement of force-distance curves. An electronic "sample and hold" circuit detects the peak force. This peak force serves as the input of the control loop and is the fed into the SFM control electronics, providing a constant force measurement. The feedback loop maintains now a constant force. The PFM automates the simultaneous measurement of topography and surface properties at every point on the sample by recording only a few characteristic points of the curves using additional electronic circuits. Data taken from the curves are the peak force (3), the local stiffness (2) and the pull-off force (4) (see Figure 2). The maximum force serves as the input of the control loop and will give the topography image. The local stiffness is measured by recording the slope of the force-distance curve subtracting the sample-hold signals at the peak value (3) and a second deflection value at a given time in the repulsive part of the curve (2). Local adhesion is detected by recording the pull-off force (4) in the force-distance curve using a peak picker circuit. The adhesion force output, that gives the adhesive force image, is the difference between the output of a peak-detector for negative forces and a "sample and hold" measuring the zero base line (1). A detailed description of this technique has been published by Rosa-Zeiser et al (13).

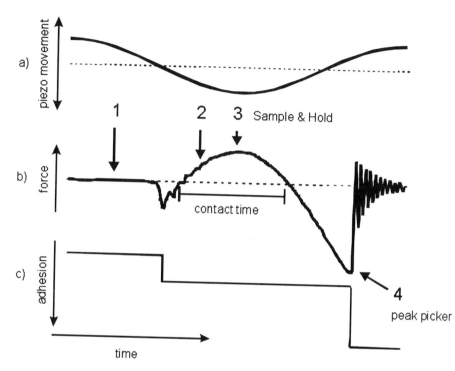

Figure 2. Schematic of the Pulsed Force Mode operation.
(a) depicts the sine wave movement of the piezo, (b) a typical change in force, and (c) the resulting adhesive force obtained through the peak picker and sample-hold circuitry. $1 = F_0$ $3 = F_{max}$ $2 = F_{is}$ $4 = F_{adh}$
Topography $= F_{max}$; Local stiffness $= F_{max} - F_{is}$; Adhesion $= F_0 - F_{adh}$

Usually, the force values measured by PFM are recorded in voltage. To quantify the adhesion data the deflection signal has to be converted into force values by the following equation:

$$F_{adh} = \frac{S * C * f}{A * s^{\bullet}}$$

(1)

where F_{adh} is the measured pull off force, S is the detected force in volt, A is an amplifying factor, which can be adjusted in the PFM-circuitry. C is a conversion factor given by the SFM system (for the system used C=6.57). The sensitivity s^{*} of the detection unit are calibrated by estimating tip-sample force curves on a bare silicon substrate. f is the spring constant of the cantilever used for the experiment.

Adhesion force experiments were performed on a TopoMetrix Explorer system equipped with the Pulsed Force Mode. For the PFM measurement either rectangular silicon or silicon nitride cantilevers (lateral force tips, Nanosensors or Olympus) with resonance frequencies between 70 and 90 kHz were used. The manufacturer's nominal force values have been used as spring constant for calculation of the adhesion forces. Images were acquired with a scan rate of 1 Hz, and a modulation frequency of 996 Hz with a 400 nm deflection. Tip-sample forces were minimized to be below 10 nN. The adhesive force values were determined from the peak values of adhesive force histograms. The accuracy of the determined force values is +/- 5 nN.

Results and Discussion

ESCA Analysis

Compositional analysis of the HFP plasma polymer surfaces reflected dominant signals arising from fluorine and carbon. Only traces of oxygen and nitrogen were detected (< 1%), which probably originated from exposure to air and the adsorption of contaminants. A distinct envelope pattern was found in high resolution analysis of the carbon C1s peak revealing peaks for $\underline{C}F_3$, $\underline{C}F_2$, $\underline{C}F$ and $\underline{C}$-CF (Figure 3). The simple monomer structure of HFP and the chosen reactor conditions results in number of different chemical species in the plasma gas. These species react with each other and form the plasma polymer on the surface. The mechanism for recombination is not known. The C1s envelope of the plasma polymer changed as a function of applied reactor energy indicating that chemical composition could be influenced. The intensities and the ratio of these peaks were dependent on the applied reactor power. The calculated F/C ratio changed from about 1.5 for 2 Watt applied reactor power to 1.3 for 20 W and increased again above 1.4 for higher reactor powers (40 W or more). That bell-shaped curve indicated a change in composition, which is discussed in more detailed in elsewhere (9). The complex C 1s envelop structure of the peaks around 286 eV indicated that species with double bond character were present. It is suggested that some of HFP monomers were partially retained or that double bonds were formed upon recombination of the reactants in the plasma polymer. This finding is consistent with recent investigations on other plasma polymers (24).

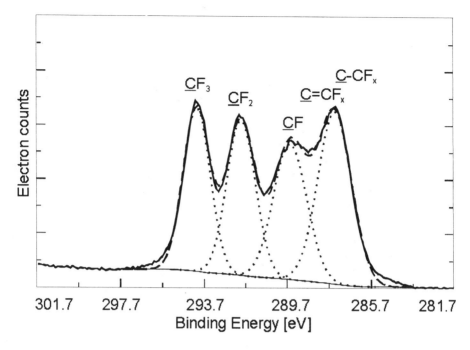

*Figure 3 High resolution ESCA spectrum of a 10 W HFP plasma polymer film
The CF₃ and CF₂ peaks are pronounced and separated from CF and C-CFₓ/C=CFₓ.
The broadening of two peaks at lower energy indicated the presents of addition
chemical species, which could not be independently resolved.*

It has been shown that HFP does not form any ordered plasma polymer structure
under the used "in glow" conditions and that the polymer is randomly crosslinked
(25). The crosslinking or branching degree can be used as to describe the plasma
polymer structure. Plasma polymers do not have a defined molecular mass as do
polymers produced by classical chemical chain reaction. High resolution ESCA
analysis allowed to distinguish chemical species in the polymer, which act either as
chain termination (e.g. CF_3), chain elongation (e.g. CF_2) or contribute to branching
and crosslinking (e.g. CF). CF_3 and CF_2 species, which do not contribute at all to
crosslinking, showed distinct peaks at the high binding energy edge (CF_3 293.8 eV
and CF_2 292 eV). It is assumed that the envelope of peaks at energies lower than 292
eV contained functional groups that contributed to crosslinking. Species such as
$FC=C$, $CF-CF$ or $C-CF$ may act as branching sites. The energy resolution of the C 1s
envelope, however, did not allow for separation of the species mentioned above from
$FC=CF_2$ or $FC=CF$, which would act as chain elongating or terminating species.
Nevertheless we did not differentiate between those species in our calculations as it is
supposed that introduced error is insignificant. The crosslinking degree was
calculated by adding up the % of the species contributing to branching and divided by
the total %.

$$crosslinking[\%] = \frac{(peak3)[\%] + (peak4)[\%]}{Total[\%]} \qquad (2)$$

The calculated crosslinking density showed a bell-shaped behavior with minimum at low and high applied reactor energies and a maximum near 20 W applied reactor energy (Figure 4b). The minimum for low applied reactor power was expected, as the reactor energy was not high enough to produce a maximal number of carbon radicals, which could contribute to branching. With higher reactor power the number of carbon radicals increases, and a higher crosslinking degree in the film will occur. The maximum was observed approximately at 20 W applied reactor power, which coincided with the lowest F/C ratio. For higher reactor energies the crosslinking decreased slightly again, which was accompanied by an increase in the F/C ratio. Due to the complexity of the plasma reaction it is not clear to us the mechanism by, which the crosslinking degree decreased for high energies as the high power radio field should be expected to produce a larger number of branching reactive species.

SFM Analysis

Film growth and thickness
The HFP films produced by RF-PECVD were very smooth and pinhole free. The RMS roughness measured in contact mode varied between 3 and 10 Å for images of 9 μm^2, 100 μm^2 respectively (9,26). The line width of the patterned samples correlated exactly with the feature size of the photomask. While no film defects were found on unpatterned fluoropolymer surfaces, on patterned samples some defects were found along the mask lines. The defect structure is introduced by the removing of the photoresist in the development process and by residue photoresist. The deposition rate of the HFP polymers was calculated from the measured film thickness of the photo lithographic patterned samples. The deposition rate increased steadily from 2 Watt to a maximum at an applied reactor powers of 20 Watts (Figure 4a). This plateau in growth rate coincided with the so-called "energy starved" range observed and described by Chen et al (7) and Garrison et al (9). For reactor powers above 20 Watt the growth rate dropped slightly. This leveling was explained by an increased "etching" of the surface at higher reactor energies. The plasma deposition process for low reactor powers was much higher than the plasma etching effect. At reactor energies above 20 W the high number of radicals slowed the deposition process down.

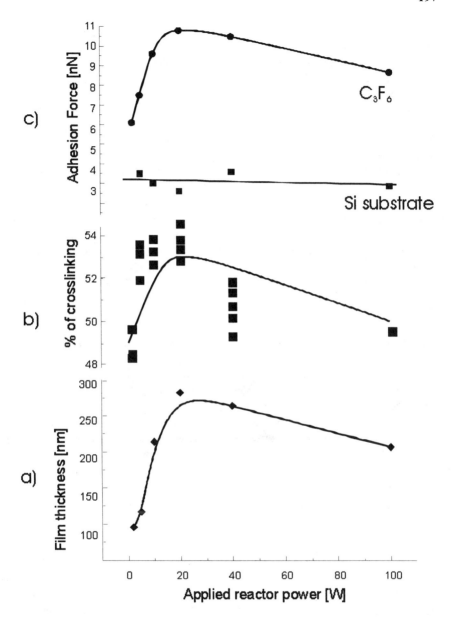

Figure 4. Analysis of HFP polymer deposited as a function of reactor power HFP plasma polymer deposited for 8 minutes as a function of applied reactor power: (a) film thickness measured by SFM, (b) calculated crosslinking and branching degrees, and (c) adhesion forces measured by PFM. The line graphs are not fitted functions and serve only as optical aid.

Adhesion force measurements

   Pulsed force imaging of the patterned plasma HFP polymer films clearly revealed differences in the tip-sample adhesion force on the fluoropolymer and silicon regions (Figure 5). Such differences in adhesion force may be due to a number of factors in polymer films including surface tension, chain mobility and contact area. For low molecular weight materials adhesion forces have been correlated to the surface tension of the material (27). If this assumption applies to the investigated system, silicon should show higher adhesive forces than the polymer because of its higher surface tension. However, we noted a 2 to 3 times higher adhesion force on the polymer films than on silicon despite the low surface energy of the HFP plasma polymers (contact angle $\Theta^A_C = 110$ °, $\Theta^R_C = 91$ ° respectively). This might be due to the high degree of crosslinking in the films that effectively increased the molecular weight. Here, it should be noted that the contact angle did not appear to be sensitive to the subtle changes in film chemistry. Measurements on thin polystyrene films have indicated that the mobility of the polymer may also influence the adhesion force due to wetting of the tip, even in polymers of higher molecular weight (28,29). Again, the high degree of crosslinking in these films would argue against high chain mobility. However, we do observe that these films are highly adherent to the silicon substrate and SFM probes coated likewise (unpublished data).

   Analyzing the adhesive force as a function of applied reactor power revealed minima at low and high powers with a maximum adhesion force around 20 W (Figure 4c). The similarity of the graphs for crosslinking degree and adhesion force suggests a direct relation. Adhesion forces depend significantly upon the contact area. Since the SFM is operated in a constant force mode, the contact area of the tip will vary as a function of the elastic modulus. Therefore, adhesion force measurements on softer materials will result in larger areas of interaction, and hence more adhesive interactions that on more stiff materials. In addition, substrate effects can not be excluded for certain. As the samples were probed in normal direction to the substrate, the compressibility of the polymer with the same elastic modulus is strongly dependent on the film thickness. Thin films are strongly influence by the substrate while thick films exhibit bulk properties. To test this argumentation additional experiments have to be carried out.

   As mentioned above, a high degree of crosslinking would argue for a higher Young's Modulus, which should result in a lower contact area and, therefore, in a lower adhesion force. However, the highest adhesion force was obtained on the highest crosslinked film. Also it was noted that the film deposition (growth) rate followed the same trend as crosslinking degree and adhesive force. This faster film growth with a concurrent increase in the density of crosslinking could result in more porous polymer films that are less dens. It seems that the deposition rate is a critical component for the elastic properties of the polymers. A slow deposition rate allows for a higher organization of the polymer chains and a lower entanglement. As the power is increased the films are faster deposited and trapped in a less organized, more mobile conformation. This situation continues until the increase applied reactor power drives the reaction into an etching regime. The competitive etching and deposition permit the reorganization of the polymer into a lower energy state. This argumentation is supported by the observation of Castner et al (25) who described highly oriented plasma polymer films slowly grow in the "after glow regions" and very recent investigations of the shear behavior and the shear modulus of those films indicated as well a higher order at slower deposition rates (11). We believe that the measured adhesion force reflect more the changes in the contact area and chain mobility than the alterations in the surface chemical composition. Both, the elastic properties and the chemical composition are dependent and controlled by the plasma deposition conditions.

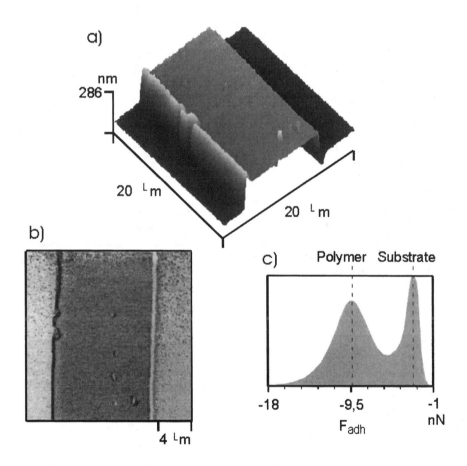

Figure 5. Patterned HFP plasma polymer film on silicon
SFM image of an 8 minutes 10 W HFP plasma film obtained in the pulsed force mode. (a) topography, (b) adhesion force, and c) corresponding scale.

Conclusions

In our work we have shown that the chemical composition and adhesive properties changed as a function of applied reactor power on RF-PECVD deposited hexafluoropropylene films. Films deposited at 20 W applied reactor energy showed the largest film thickness, the highest crosslinking degree and the highest adhesion force. Those parameters showed a logarithmic increase up to 20 W applied reactor power. At higher reactor power values, a linear decrease or slightly logarithmic decay (Bradley model) was observed. The crosslinking degree as determined via ESCA was calculated to be in a narrow bandwidth between 48 % and 54 %. Nevertheless, the differences were enough to give a significant change on the different film types.

The deposition rate, and therefore the film thickness, changed dependent on the applied reactor power. However, the film roughness did not show any dependency on deposition rate. It ranged between 0.3 nm and 1 nm, which was in the same range as the substrate roughness of silicon. We conclude that the film formation occurred at the substrate surface, and that limited polymerization of hexafluoropropylene took place in gas phase.

The variable reactor conditions not only influenced the chemical composition and the deposition rate, but were reflected in a change of adhesion force as well. The measured adhesion forces between the SFM tip and the substrate ranged between 6 nN and 11 nN with the maximum at 20 W applied reactor energy while the adhesion force on silicon was at about 3.5 nN. We explain this large variation in adhesive force as a change in contact area, and therefore, a change in Young's modulus. The molecular structure of the HFP plasma polymers were influenced by the changed deposition rate. Faster deposition rates resulted in more porous, less organized polymers.

By controlling the reactor conditions we can tailor film chemistry and the micro and nano mechanical properties. In order to better understand the change in structure of the plasma HFP polymer films further investigations on the nano contact mechanical properties have to be carried out.

Acknowledgments

This work was supported by UWEB (University of Washington Engineered Biomaterials), the National ESCA and Surface Analysis Center for Biomedical Problems (NESAC/BIO NIH NCRR grant RR01296) and the Whitaker Foundation. Reto Luginbühl was supported by a Swiss National Science Fellowship, Mike Garrison through a US Department of Education GAANN Fellowship and the University of Washington Department of Bioengineering. René Overney was supported in part by the Royal Research Foundation (UW) and the Exxon Educational Foundation, and Sabine Hild by Margarethe-von-Wrangell-Stiftung. Special thanks to Jason Christos, Winston Ciridon for their help in preparing samples and Othmar Marti, Dave Castner and Andreas Goessel for their stimulating discussions.

References

1) Johnston E. E.; Bryers J. D.; Ratner B. D., *ACSPolymer Preprints* **1997**, *38*, 1016.

2) Kiaei D.; Hoffman A. S.; Hanson S. R., *J Biomed Mater Res* **1992**, *26*, 357.

3) Kiaei D.; Hoffman A. S.; Horbett T. A.; Lew K. R., *J Biomed Mater Res* **1995**, *29*, 729.

4) Favia P.; Perez-Luna V. H.; Boland T.; Castner D. G.; Ratner B. D., *Plasmas and Polymers* **1996**, *1*, 299.

5) Kiaei D.; Hoffman A. S.; Horbett T. A., *J Biomater Sci Polym Ed* **1992**, *4*, 35.

6) Silverstein M. S.; Chen R.; Kesler O., *Polymer Engineering and Science* **1996**, *36*, 2542.

7) Chen R.; Gorelik V.; Silverstein M. S., *Journal Of Polymer Science* **1995**, *56*, 615.

8) Luginbühl R.; Garrison M. D.; Overney R. M.; Weiss L.; Hild S.; Ratner B. D., *Polymer Preprints* **1998**, *39*, 932.

9) Garrison M. D.; Luginbühl R.; Overney R. M.; Ratner B. D., *Thin Solid Films* **1999**, *352*, 13.

10) Goessel A.; Lhoest J.-B.; Bowen-Pope D.; Hoffman A. S. *Plasma Lithography - Heterogeneous Polymer Substrates for the Patterned Immobilization of Ligands*: Seattle, 1998.

11) Luginbühl R.; Garrison M. D.; Overney R. M.; Ratner B. D., *in preparation* **1999**.

12) Radmacher M.; Tillmann R. W.; Gaub H. E., *Biophys.J.* **1993**, *64*, 735.

13) Rosa-Zeiser A.; Weilandt E.; Hild S.; Marti O., *Measurement Science & Technology* **1997**, *8*, 1333.

14) Garrison M. D.; Luginbühl R.; Golledge S.; Christos J.; Goessl A.; Ratner B. D., *in prepartion* **1999**.

15) Morosoff N.,*An Introduction to Plasma Polymerization*; d'Agostino, R., Ed.; Academic Press Inc.: San Diego and London, 1990, pp 1.

16) Chen R.; Silverstein M. S., *Journal Of Polymer Science Part a-Polymer Chemistry* **1996**, *34*, 207.

17) Charlson E. J.; Charlson E. M.; Sharma A. K.; Yasuda H. K., *J. Appl. Polym. Sci. Appl. Polym. Symp.* **1984**, *38*, 137.

18) Diks A.,*Theory, Techniques and Applications*; Baker, A. D. and Brundle, C. R., Ed.; Academic Press: London, 1981, pp 277.

19) Binnig G.; Quate C. F.; Gerber C., *Physical Review Letters* **1986**, *56 (9)*, 930.

20) Mizes H. A.; Loh K. G.; Miller R. J. D.; Ahuja S. K.; Grabowski E. F., *Applied Physics Letters* **1991**, *59*, 2901.

21) van der Werf K. O.; Putman C. A. J.; de Grooth B. G.; Greve J., *Applied Physics Letters* **1994**, *65*, 1195.

22) Marti O.,; Bhushan, B., Ed.; Kluwer Scientific Publishers: Dordrecht, 1996; Vol. E:330, pp 17.

23) Marti O.; Hild S.; Staud J.; Rosa A.; Zink B.,; Bhushan, B., Ed.; Kluwer Scientific Publishers: Dordrecht, 1996; Vol. E:330, pp 455.

24) Pan Y. V.; Wesely J.; Denton D.; Ratner B. D., *in preparation* **1999**.
25) Castner D. G.; Lewis K. B.; Fischer D. A.; Ratner B. D.; Gland J. L., *Langmuir* **1993**, *9*, 537.
26) Christos J.; Luginbuehl R.; Pan Y. V.; Ratner B. D., *Journal of Undergraduate Research in Engineering* **1998**, *2*, 52.
27) Berger C. E. H.; van der Werf K. O.; Kooyman R. P. H.; Degrooth B. G.; Greve J., *Langmuir* **1995**, *11*, 4188.
28) Hild S.; Marti S., *Polymer Preprints* **1998**, *39*, 1230.
29) Marti O.; Waschipky H.; Stifter T.; Quintus M.; Hild S., *Surfaces and Colloids A* **1998**, *submitted*.

Chapter 15

Fluoropolymer Films Deposited by rf Plasma Sputtering of Polytetrafluoroethylene Using Inert Gases

M. A. Golub[1,3] and T. Wydeven[2]

[1]NASA Ames Research Center and [2]Lockheed Martin Engineering and Sciences, Moffett Field, CA 94035–1000

The FT-IR, UV and XPS spectra of fluoropolymer films (SPTFE) deposited by rf plasma sputtering of polytetrafluoroethylene (PTFE), using Ne, Kr and Xe as sputtering gases, were obtained and compared with prior spectra for SPTFE formed using He and Ar. The F/C ratios for SPTFE films (1.44–1.55), obtained at an rf power of 10 W, were essentially the same for all five rare gases, with no trend of decreasing fluorine content in the SPTFE product with increasing atomic weight of the sputtering gas. Increasing rf power from 10 to 50 W resulted in successively lower F/C ratios for SPTFE (e.g., from 1.55 to 1.21 in the case of Xe plasma-sputtered PTFE), accompanied by sputtering of the Pyrex glass reactor and deposition of elements of the glass occurring at 40 W and above. In order to achieve a "Teflon-like" SPTFE structure (i.e., products with as high an F/C ratio as possible) in a given plasma reactor, an optimum rf power must be found, which in the present case was approximately 10 W.

The FT-IR, XPS and UV spectra of plasma-polymerized tetrafluoroethylene (PPTFE) and of the fluoropolymer deposits (SPTFE) formed from rf plasma sputtering of polytetrafluoroethylene (PTFE) using He or Ar as sputtering gas were reported previously (1). In apparently the first study to involve preparation of PPTFE and SPTFE deposits in the same reactor and under comparable low-power (10 W) plasma conditions, the general similarity of PPTFE and SPTFE microstructures—noted in prior literature (2-5) on the basis of their IR and/or XPS spectra—has been reconfirmed, with some differences observed. Inasmuch as tetrafluoroethylene (TFE) monomer is the main product formed in the He plasma-induced decomposition of PTFE (6), it is not surprising that in-situ generation of TFE in rf plasma sputtering of PTFE yielded polymeric deposits that were similar to those obtained by plasma polymerization of TFE.

From C_{1s} XPS spectra, the average values and standard deviations for the fluorine-to-carbon (F/C) ratios for a series of SPTFE deposits formed using either He or Ar (SPTFE-H or -A) as the sputtering gas were 1.56 ± 0.03 and 1.49 ± 0.04, respectively (1), in contrast to the theoretical value of 2.0 for pristine PTFE. These F/C ratios are within the range of literature values (1.3-1.6) (5,7-10) for SPTFE, and

---

[3]Current address: 12,000 4th Street North, #205, St. Petersburg, FL 33716 (retired)

also comparable to F/C ratios (1.2-1.6) for PPTFE (*1*). However, the F/C ratios for SPTFE-H and -A are distinctly higher than the values reported by Hishmeh et al. (*11*) for SPTFE generated in an Ar plasma (0.89), and by Ryan et al. (*12*) for SPTFE deposits formed in He, Ne and Ar glow discharges (1.02, 0.82 and 0.78, respectively). The latter workers attributed the progressive drop in F/C ratio in moving from He to Ne to Ar to momentum transfer phenomena related to the different atomic masses of the impinging rare gases.

Since the difference between F/C ratios for Ryan et al.'s SPTFE deposits obtained by He and Ar sputtering (0.24) was much greater than the corresponding difference for our SPTFE-H and -A deposits (0.07), we were prompted to extend our spectroscopic study of SPTFE (*1*) to include three other rare gases (Ne, Kr and Xe) as sputtering agent, to test the momentum transfer concept. Moreover, since Ryan et al. carried out their rf plasma sputtering at a power of 50 W, while ours was done at 10 W, the new work with Ne, Kr and Xe included some runs at 20 to 50 W in addition to the desired, comparative runs at 10 W. In this way, we aimed to examine also the effect of rf power on the F/C ratios for SPTFE and, at the same time, try to account for the striking differences noted between the respective FT-IR and C_{1s} XPS spectra, as well as F/C ratios, for SPTFE products in our prior work (*1*) and in that of Ryan et al. (*12*).

Experimental

The apparatus employed for rf plasma sputtering of PTFE film (0.076 mm thick; Chemfab), using high-purity Ne, Kr or Xe as the sputtering agent, and the analytical equipment were the same as those used previously (*1*). A fresh, cleaned PTFE target sheet (25 cm × 10 cm) was used in each sputtering run and deployed as described earlier (*1*). KBr disks, small pieces of Si wafers and quartz windows served as substrates for FT-IR, XPS and UV spectral analyses of SPTFE deposits. XPS spectra were obtained with an SSX-100 spectrometer at Surface Science Laboratories, Mountain View, CA. While most of the sputtering runs were carried out with an rf power of 10 Watts, several runs with Ne, Kr and Xe were also conducted at 20, 30, 40 and 50 Watts. The initial pressure (25 mTorr) and flow rate (0.5 cm^3 STP/min) of the rare gas were the same in all runs, with sputter deposition times of up to 2 h.

Results and Discussion

Rf plasma sputtering at 10 Watts. Figure 1 compares typical C_{1s} XPS spectra of SPTFE formed using Ne (SPTFE-N), Kr (SPTFE-K) or Xe (SPTFE-X) as the sputtering gas. These spectra are not only virtually identical in shape or appearance, they are also essentially indistinguishable from the C_{1s} XPS spectra of SPTFE-H and -A reported previously (Fig. 2 in ref. 1). In common with the latter spectra, those of SPTFE-N, -K and -X exhibited four prominent "finger-like" peaks having the following, customary assignments (*1*): 294.0-294.1 ($-CF_3$), 292.0 ($-CF_2-$), 290 (>CF-; -CF=), and 287.8 eV (≯CCF<; ≯CO-), along with a very weak, fifth peak at 285.3-285.5 eV that represents a trace of hydrocarbon ($-CH_2-$) surface contamination. Average values and standard deviations for the F/C ratios for SPTFE-N, -K and -X, as calculated from the relative intensities of the deconvoluted peaks in the C_{1s} spectra (*1*), came to 1.48 ± 0.07, 1.44 ± 0.05 and 1.55 ± 0.01, respectively, which are comparable to the corresponding values of F/C for SPTFE-H and -A (1.56 ± 0.03 and 1.49 ± 0.04, respectively) given above. The F/C values for all five rare gases may be regarded as virtually the same, there being no trend of decreasing F content in SPTFE product with increasing atomic weight of the sputtering gas at 10 W, contrary

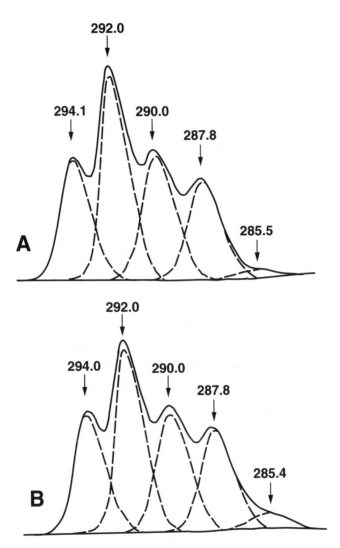

Figure 1. Typical C_{1s} XPS spectra of (A) SPTFE-X, (B) SPTFE-K, and (C) SPTFE-N formed at 10 W. Binding energies are in electron volt (eV) units.

Continued on next page.

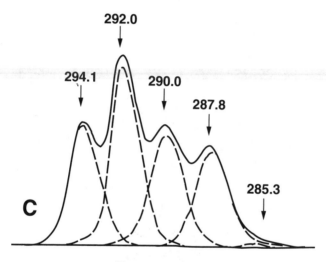

Figure 1. *Continued.*

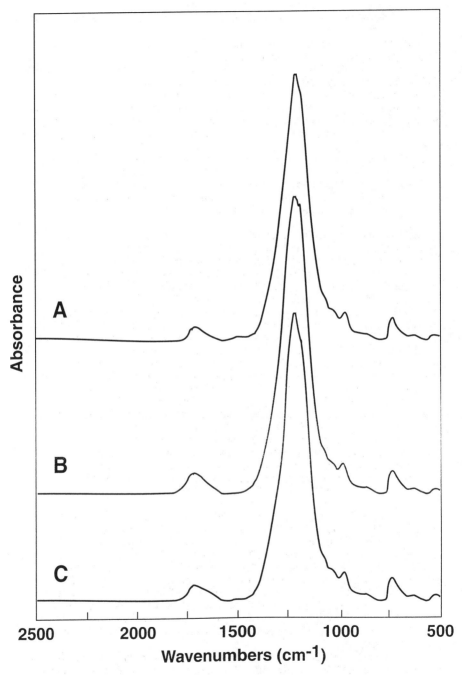

Figure 2. Typical FT-IR spectra of (A) SPTFE-X, (B) SPTFE-K, and (C) SPTFE-N formed at 10 W.

to the momentum transfer idea and data reported by Ryan et al. (*12*) for sputtering at 50 W.

As with the C_{1s} XPS spectra, so also are the FT-IR spectra of SPTFE-X, -K and -N at once identical to, and indistinguishable from, the FT-IR spectra of SPTFE-A and -H reported previously (Fig. 1 of ref. 1). The former FT-IR spectra (Fig. 2) likewise exhibited a characteristic intense, broad absorption band centered at ca. 1250 cm^{-1} (overlap of CF, CF_2 and CF_3 vibrations), plus weak absorptions at ca. 750 cm^{-1} ("amorphous" $-CF_2-$ structure) and at ca. 1720 cm^{-1} ($-CF=CF-$ and $>C=O$ groups). Interestingly, those spectra lacked the strong, broad (and extraneous) absorptions in the 1400-1800-cm^{-1} range seen in the FT-IR spectra of Ryan et al.'s SPTFE (Fig. 3 in ref. 13).

Consistent with the nearly identical FT-IR and C_{1s} XPS spectra for SPTFE-N, -K and -X were the superimposable UV spectra of these materials having similar thickness (Fig. 3). At 350 nm, the transmittance for such SPTFE films (ca. 430 nm thick) was ca. 82%. The smooth, band-free nature of the UV-visible absorption curves is indicative of conjugated double bonds (*13*).

Rf Plasma Sputtering at 20-50 Watts. Figure 4 presents typical C_{1s} XPS spectra of SPTFE-X formed at 20 and at 50 W. Increasing rf power from 20 to 50 W has clearly altered the distribution of the functionalities CF_n in SPTFE-X: intensities of the peaks at ca. 287.5 eV ($>CCF<$; $>CO-$) have increased considerably, while the intensities of the peaks at 294.0 eV ($-CF_3$) and at ca. 289.8 eV ($>CF-$; $-CF=$) have increased slightly, relative to that of the major peak at 292.0 eV ($-CF_2-$). The peaks at ca. 285.3 eV (adventitious hydrocarbon) have, surprisingly, also increased in relative intensity with increase in rf power. The C_{1s} XPS spectra of SPTFE-X formed at 30 and 40 W (not shown here) had shapes intermediate between those of Spectra A and B in Figure 4, while Spectrum A in this figure (for 20 W) is intermediate between that of SPTFE-X formed at 10 W (Spectrum A in Fig. 1) and that of SPTFE-X formed at 30 W. These spectral changes reflect a net, progressive decline in F content with increase in rf power: the F/C ratios for SPTFE-X at the different powers were 1.55 (10 W); 1.49 (20 W); 1.39 (30 W); 1.34 (40 W) and 1.21 (50 W). A similar effect of rf power on F/C ratio was observed in this work for SPTFE-N and -K, and retrospectively (*1*) for SPTFE-H and -A.

At 40 W and above, sputtering of the Pyrex glass reactor also occurred, giving rise to simultaneous deposition of elements of glass along with SPTFE. Thus, the XPS survey scans for SPTFE-X formed at different powers revealed the presence of the following extraneous elements besides the expected C and F: O, 0.5–0.7 atom % (10–30 W); Na, 1.2%; O, 1.3% (40 W) and Na, 6.3%; Si, 0.9%; Al, 0.9%; O, 3.2% (50 W). Somewhere between 40 and 50 W, a further complication was enhanced ablation of the SPTFE, resulting in extremely low rates of SPTFE formation (and hence especially thin deposits with poor FT-IR spectra) at 50 W, but respectable rates at ≤40 W. (At 50 W, but not at 40 W, the 0.076-mm PTFE target was partially disintegrated or etched away after ca. 2 h of sputtering.) The FT-IR spectra of SPTFE-N, -K and -X products formed at 20–40 W (not shown here) resembled closely the FT-IR spectra for the corresponding products formed at 10 W (Fig. 2); however, the FT-IR spectra for the glass-contaminated SPTFE formed at 50 W (also not shown here) revealed absorptions in the 1000-1300-cm^{-1} range indicative of both Si-O and C-F functionalities, but none of the extraneous absorptions (at 1400–1800 cm^{-1}) found in the FT-IR of Ryan et al.'s Ar plasma-sputtered PTFE (*12*) which was also formed at 50 W.

An important point to note is that for a given plasma reactor there is an optimum rf power for producing "Teflon-like" SPTFE products (i.e., products with as high an F/C ratio as possible); in our case, that power is about 10 W. Interestingly, the F/C ratio for SPTFE-X formed at 50 W (1.21) is still larger than the F/C ratio for Ryan et

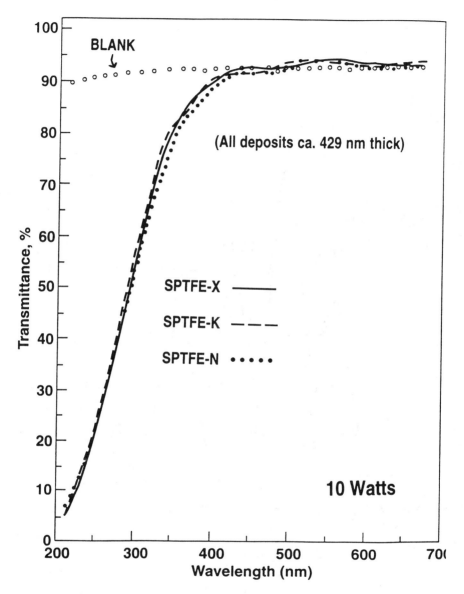

Figure 3. Typical UV-visible spectra of SPTFE-X, -K, and -N deposits (all ca. 429 nm thick) and of the uncoated (blank) quartz window.

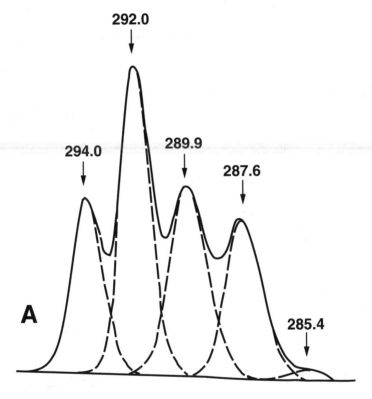

Figure 4. Typical C_{1s} XPS spectra of SPTFE-X formed at 20 W (A) and at 50 W (B).

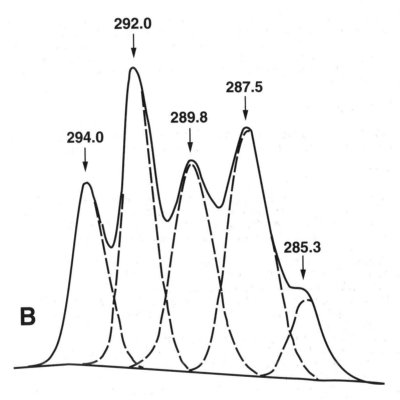

Figure 4. *Continued.*

al.'s (*12*) Ar plasma-sputtered PTFE (0.89) formed at the same power. In trying to reconcile our results for SPTFE (here and in the prior work (*1*)) with those of Ryan et al., we consider their products to have been subjected to excessive defluorination (perhaps in their use of a nonmonochromatized MgKα X-ray source for XPS) and/or contamination (perhaps through concurrent sputtering of some unknown material in their reactor). At any rate, the present work demonstrates that SPTFE structures that retain a high percentage of the F content present in the original PTFE may be obtained by rf plasma sputtering with any of the rare gases.

Conclusions

The F/C ratios for SPTFE prepared at 10 W (1.44–1.55) were essentially the same whether He, Ne, Ar, Kr or Xe was used as the sputtering agent. Increasing rf power from 10 to 50 W for rf plasma-sputtering of PTFE resulted in successively lower F/C ratios (e.g., from 1.55 to 1.21 in the case of SPTFE-X), accompanied by sputtering of the Pyrex glass reactor and deposition of glass-decomposition products occurring at 40 W and above.

Acknowledgment

The authors express appreciation to Jeanie Howard, Lockheed Martin Engineering and Sciences, for providing the FT-IR and UV spectra for the various SPTFE samples.

Literature Cited

1. Golub, M. A.; Wydeven, T.; Johnson, A. L. *Langmuir* **1998**, *14*, 2217.
2. Dilks, A.; Kay, E. *Macromolecules* **1981**, *14*, 855.
3. Tibbitt, J. M.; Shen, M.; Bell, A. T. *Thin Solid Films* **1975**, *29*, L43.
4. Yasuda, H. *Plasma Polymerization*; Academic Press; Orlando, FL, 1985; pp 184-185.
5. Yamada, Y.; Kurobe, T. *Jpn. J. Appl. Phys.* **1993**, *32*, 5090.
6. Mathias, E.; Miller, G. H. *J. Phys. Chem.* **1967**, *71*, 2671
7. Horie, M. *J. Vac. Sci. Technol. A* **1995**, *13*, 2490.
8. Pireaux, J. J.; Delrue, J. P.; Hecq, A.; Duchot, J. P. In *Physico-chemical Aspects of Polymer Surfaces*; Mittal, K. L., Ed.; Plenum Press: New York, 1984; pp 53-81.
9. Sugimoto, I.; Miyake, S. *J. Appl. Phys.* **1988**, *64*, 2700.
10. Maréchal, N.; Pauleau, Y. *J. Vac. Sci. Technol.* **1992**, *A10*, 477.
11. Hishmeh, G. A.; Barr, T. L.; Sklyarov, A.; Hardcastle, S. *J. Vac. Sci. Technol.* **1996**, *A14*, 1330.
12. Ryan, M. E.; Fonseca, J. L. C.; Tasker, S.; Badyal, J. P. S. *J. Phys. Chem.* **1995**, *99*, 7060.
13. Sugimoto, I. *Macromolecules* **1991**, *24*, 1480.

Chapter 16

Fluoropolymer Films Deposited by Argon Ion-Beam Sputtering of Polytetrafluoroethylene

M. A. Golub[1,4], B. A. Banks[2], S. K. Rutledge[2], and M. C. Kitral[3]

[1]NASA Ames Research Center, Moffett Field, CA 94035–1000
[2]NASA Lewis Research Center, Cleveland, OH 44135
[3]Cleveland State University, Cleveland, OH 44115

The FT-IR, XPS and UV spectra of fluoropolymer films (SPTFE-I) deposited by argon ion-beam sputtering of polytetrafluoroethylene (PTFE) were obtained and compared with prior corresponding spectra of fluoropolymer films (SPTFE-P) deposited by argon rf plasma sputtering of PTFE. Although the F/C ratios for SPTFE-I and -P (1.63 and 1.51) were similar, the SPTFE-I structure had a much higher concentration of CF_2 groups than the SPTFE-P structure: ca. 61 and 33% of the total carbon contents, respectively. Reflecting that difference, the FT-IR spectra of SPTFE-I showed a distinct doublet at 1210 and 1150 cm^{-1} whereas SPTFE-P presented a broad, featureless band at ca. 1250 cm^{-1}. The absorbance of the 1210-cm^{-1} band in SPTFE-I was proportional to film thickness in the range of 50–400 nm. SPTFE-I was more transparent in the UV than SPTFE-P at comparable thickness. The mechanism for SPTFE-I formation likely involves "chopping off" of oligomeric segments of PTFE as an accompaniment to "plasma" polymerization of TFE monomer generated in situ from PTFE on impact with energetic Ar ions. Data are given for SPTFE-I deposits and the associated Ar^+-bombarded PTFE targets where a fresh target was used for each run or a single target was used for a sequence of runs.

Argon ion-beam sputtering of polytetrafluoroethylene (PTFE) has attracted considerable research interest from the standpoint of the surface modification of the PTFE target (1-5) or the formation of sputter-deposited, fluoropolymer coatings on various substrates (6-10). Regarding surface modification, a strong similarity between the microstructural changes produced in PTFE when subjected to either a low-energy Ar+ ion beam or an Ar plasma had been noted by Wells et al. (3), and later validated (5) by the related work of Tan et al. (2) and Golub et al. (11). Regarding sputter-deposited fluoropolymers, one of us reported recently on the microstructural similarity between plasma-polymerized tetrafluoroethylene (PPTFE) and the fluoropolymer deposits formed from rf plasma sputtering of PTFE (SPTFE-P), first using He or Ar as sputtering gas (12), and subsequently using three other inert gases (Ne, Kr and Xe) as sputtering gases (13). Having briefly examined some years ago fluoropolymer deposits formed from Ar ion-beam sputtering of PTFE (SPTFE-I) (7,8), we were prompted to carry out a more complete examination of SPTFE-I using FT-IR, XPS and UV-visible spectroscopy for

---

[3]Current address: 12,000 4th Street North, #205, St. Petersburg, FL 33716 (retired)

microstructural characterization as well as for comparison of SPTFE-I and SPTFE-P products.

Experimental

The ion-beam sputtering technique peformed was described previously (7) and utilized ion-beam sputter cleaning of the target and deposition surfaces prior to sputter deposition. The PTFE targets used in this study were 12.7-cm disks of DuPont Teflon located 21.3 cm downstream of a 2.5-cm beam diameter Ar ion source used for ion-beam sputtering. The sputter target was inclined 45° from perpendicular to the ion beam. The deposition surfaces were located 17.1 cm from the sputter target and perpendicular to a line-of-sight between them. Typically, a separate PTFE target was used in each sputtering run, but some runs involved sequential sputtering of a single target. KBr disks (for FT-IR), Si wafers (for XPS), and quartz disks (for UV analyses of SPTFE–I deposits) were used as substrates in the deposition. The Ar ion-beam conditions used to prepare the SPTFE deposits were: Ion energy, 750 eV; ion beam current, 28 mA; chamber pressure, 0.1 mTorr; sputter-deposition times, 2.5-51.0 min. FT-IR absorption spectra of SPTFE were obtained by means of a Mattson Model 6020 spectrometer; UV-visible spectra were obtained with a Cary 3 spectrophotometer, while corresponding XPS spectra were obtained with an SSX-101 spectrometer at Surface Science Laboratories, Mountain View, Calif., using monochromatized Al Kα X-rays, the binding energies being referenced to the F_{1s} peak at 689.2 eV.

Results and Discussion

Figure 1 compares typical C_{1s} XPS spectra of the SPTFE deposits formed by Ar ion-beam sputtering of PTFE (SPTFE-I) in this study and by Ar rf plasma sputtering of PTFE (SPTFE-P) as reported previously (12). Both spectra display four characteristic peaks having the following customary assignments (14): 294.0-294.1 (-CF$_3$), 292.0 (-CF$_2$-), 289.8-289.9 (>CF-; -CF=) and 287.6-287.7 eV (≥CCF-; ≥CO-), along with a weak, fifth peak at 285.0-285.4 eV that represents hydrocarbon surface contamination (-CH$_2$-) and/or crosslinks. The F/C ratios for SPTFE-I and -P, as calculated (12) from the relative intensities of the deconvoluted peaks in the C_{1s} spectra, were 1.63 and 1.51, respectively. The small difference between these F/C ratios belies, however, an important structural difference between SPTFE-I and -P, namely, a much higher concentration of CF$_2$ functionalities in the former material than in the latter (ca. 61 and 33% of the total carbon contents, respectively). Indeed, the C_{1s} spectrum of SPTFE-I resembles a composite of the corresponding spectra of PTFE (with a single -CF$_2$- peak at 292.0 eV) and of SPTFE-P, or even PPTFE (8) (with their four "finger-like" peaks). In this respect, the C_{1s} spectrum of SPTFE-I in Figure 1 (obtained with 750-eV Ar ions) resembles closely that of an earlier SPTFE-I (8) (obtained with 1000-eV Ar ions) which had an F/C ratio of 1.73 and in which the CF$_2$ groups represented ca. 70% of the total carbon content. That the prior SPTFE-I (8) had a higher F/C ratio and higher CF$_2$ content than the present SPTFE-I is in qualitative accord with the results obtained by Quaranta et al. (9) who reported SPTFE-I formed with 700- and 1000-eV Ar ions to have F/C ratios of 1.48 and 1.83, and CF$_2$ contents of 44 and 74%, respectively.

Figure 2 compares typical FT-IR spectra of SPTFE-I and -P corresponding to the respective C_{1s} spectra presented in Figure 1. In contrast to the broad featureless absorption band centered at ca. 1250 cm^{-1} (an overlap of CF, CF$_2$ and CF$_3$ vibrations) in the FT-IR spectrum of SPTFE-P, there is a distinct doublet centered at the same location in the spectrum of SPTFE-I, the doublet at 1210 and 1150 cm^{-1} being assignable to the asymmetrical and symmetrical CF$_2$ stretching modes, respectively, as seen in the FT-IR spectrum of PTFE itself. The same doublet was noted by Quaranta et al. (9), while similar doublets were seen at 1197 and 1154 cm^{-1} in the FT-IR spectra of

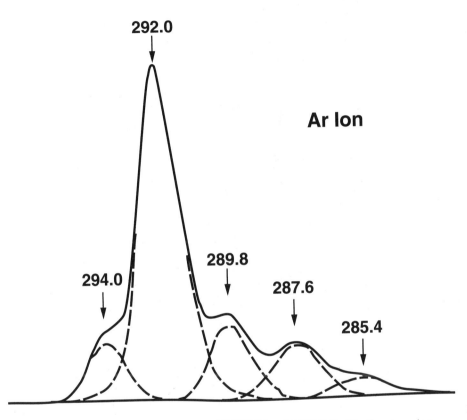

Figure 1. Typical C_{1s} XPS spectra of SPTFE-I and SPTFE-P. Binding energies are given in electron volt (eV) units.

Continued on next page.

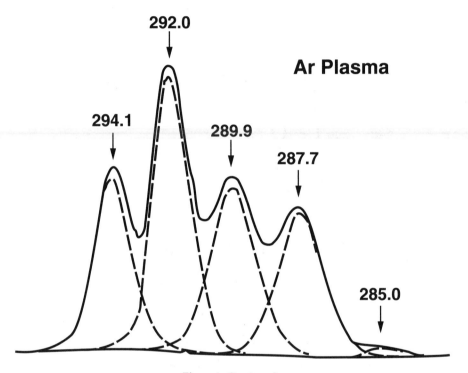

Figure 1. *Continued.*

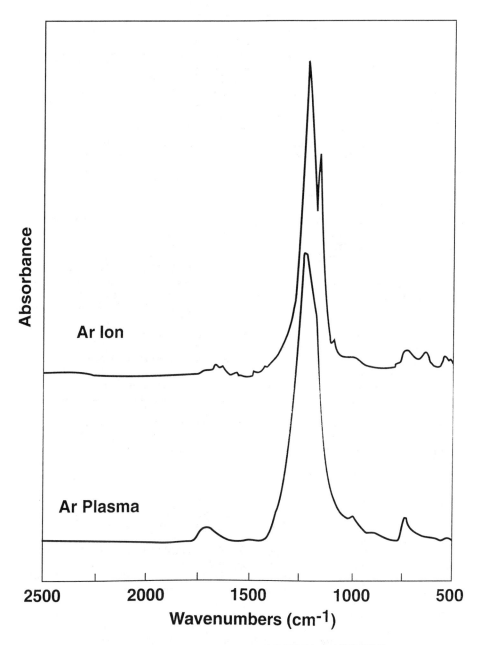

Figure 2. Typical FT-IR spectra of SPTFE-I and SPTFE-P.

Ar$^+$ sputter-deposited films from PTFE recently reported by Li et al. (10), and at 1211 and 1152 cm^{-1} in the FT-IR spectra of pulsed-laser deposited films from PTFE reported by Norton et al. (15). The presence of a distinct doublet in the FT-IR spectrum of SPTFE-I, and its absence in that of SPTFE-P, reinforces the view from XPS that the former deposit is more PTFE-like in structure than the latter. Apart from the intense bands at ca. 1200–1250 cm^{-1}, there are weak absorptions at ca. 1720 cm^{-1} (–CF=CF– and >C=O groups) and at ca. 750 cm^{-1} ("amorphous" –CF$_2$– structure). The absorbance, A, of the band at 1210 cm^{-1} for a series of SPTFE-I deposits was proportional to film thickness, t, in the range of 50–400 nm (Figure 3), and is given by the linear equation $A = 4.68 \times 10^{-4} t - 1.08 \times 10^{-3}$.

Figure 4 presents UV-visible spectra for SPTFE-I deposits (on quartz) with three different thicknesses as well as that for SPTFE-P at approximately one of those thicknesses. As may be seen, there is a steady increase in UV-visible absorption for SPTFE-I with increase in thickness, but the C$_{1s}$ spectra of the corresponding *surfaces* are all more or less alike with nearly constant F/C ratios (1.56–1.63). Interestingly, SPTFE-I is clearly more transparent than SPTFE-P for deposits of comparable thickness (353 and 359 nm, respectively). Since the smooth, band-free nature of the UV-visible spectra is indicative of conjugated double bonds (16), these UV spectral results further support the idea that the SPTFE-I structure is closer to that of PTFE than is the SPTFE-P structure.

The SPTFE-I results reported thus far have involved the use of a separate PTFE target in each sputtering run. We now consider some *limited* F/C data for SPTFE-I deposits and the associated Ar$^+$-bombarded PTFE targets where a fresh target was used for each run or a single target was used for a sequence of runs. In one experiment, a single PTFE target sputtered sequentially to yield three successive SPTFE-I deposits of increasing thickness ended up with an F/C ratio (1.88) that was somewhat less than the theoretical value (2.0) for the Ar$^+$-cleaned PTFE target, while the F/C ratios (and thicknesses in nm) for the corresponding SPTFE-I deposits were 1.24 (15); 1.47 (31); and 1.62 (149). The F/C ratio for SPTFE-I evidently increased with thickness up to a limiting value, suggesting that incorporation of F into the growing deposit increases until the concentration of F in the sputtering apparatus attains an equilibrium value. Significantly, the F/C ratio for the 149-nm thick deposit was within the range indicated above for the bulk of SPTFE-I deposits having thicknesses > 100 nm. Another experiment involved sputtering four individual PTFE targets for different periods of time to produce four SPTFE-I deposits of increasing thickness. Here, the F/C ratios for the Ar$^+$-bombarded *targets* were substantially constant (1.75, 1.53, 1.83 and 1.64), while the F/C ratios (and thicknesses in nm) for the corresponding *deposits* were 1.16 (15); 1.27 (30); 1.48 (49) and 1.57 (126). Again, the F/C ratio for SPTFE-I increased with thickness and then levelled off. Our finding that the F/C ratio for the PTFE target remained largely unchanged with time of Ar$^+$ bombardment runs counter to the observations in refs. 1 and 3 wherein defluorination of PTFE continued to very low F/C ratios (0.2 and 0.6, respectively), with no indications of any levelling-off after extended Ar$^+$ bombardment.

As suggested earlier (12), the mechanism for SPTFE-I formation likely results from a relatively large involvement of "chopping off" of oligomeric segments of PTFE macromolecules as an accompaniment to the "plasma" polymerization of TFE monomer (or other fluorocarbon fragments) generated in situ from PTFE on impact with energetic Ar ions. This view is consistent with the appearance of the C$_{1s}$ XPS spectra of SPTFE-I which resemble superpositions of the corresponding spectra for PTFE and PPTFE and which exhibit the dominant 292.0-eV peak due to CF$_2$ groups. A similar mechanism may be considered for SPTFE-P but one where the "chopping off" of oligomeric segments of PTFE macromolecules is less important than in the case of SPTFE-I.

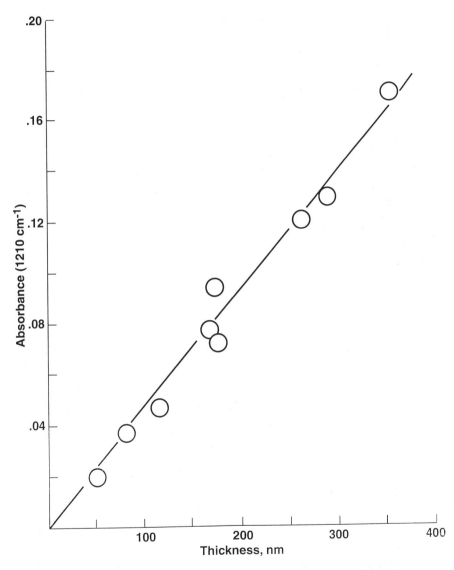

Figure 3. Absorbance of FT-IR band at 1210 cm^{-1} as a function of film thickness.

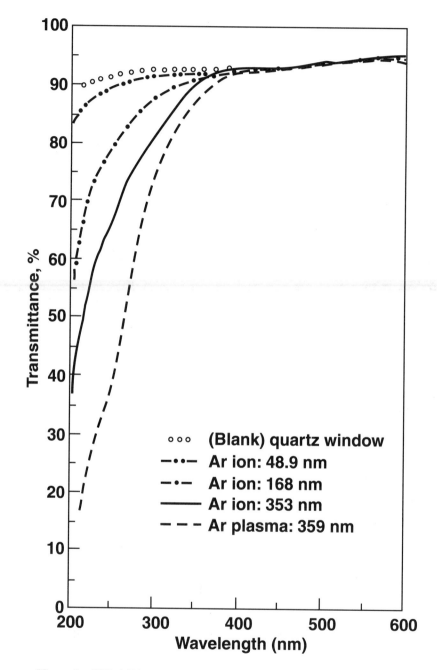

Figure 4.   UV-visible spectra for SPTFE-I (with different thicknesses) and SPTFE-P.

Acknowledgment

The authors express appreciation to Jeanie Howard, Lockheed Martin Engineering and Sciences, for providing the FT-IR and UV spectra for the various SPTFE samples.

Literature Cited

1. Takahagi, T.; Ishitani, A. *Macromolecules* **1987**, *20*, 404.
2. Tan, B. J.; Fessehaie, M.; Suib, S. L. *Langmuir* **1993**, *9*, 740.
3. Wells, R. K.; Ryan, M. E.; Badyal, J. P. S. *J. Phys. Chem.* **1993**, *97*, 12879.
4. O'Keefe, M. J.; Rigsbee, J. M. *J. Appl. Polym. Sci.* **1994**, *53*, 1631.
5. Golub, M. A. *Langmuir* **1996**, *12*, 3360.
6. Sovey, J. S. *J. Vac. Sci. Technol.* **1979**, *16*, 813.
7. Banks, B. A.; Mirtich, M. J.; Rutledge, S. K.; Swec, D. M. *Thin Solid Films.* **1985**, *127*, 107
8. Wydeven, T.; Golub, M. A.; Lerner, N. R. *J. Appl. Polym. Sci.* **1989**, *37*, 3343.
9. Quaranta, F.; Valentini, A.; Favia, P.; Lamendola, R.; d'Agostino, R. *Appl. Phys.* **1993**, *63*, 10.
10. Li, Y.; Gong, Z.; Long, Z.; Ji, C. *Dalian Ligong Daxue Xuebao* **1996**, *36*, 279; *Chem. Abst.* **1997**, *126*, 90730.
11. Golub, M. A.; Lopata, E. S.; Finney, L. S. *Langmuir* **1994**, *10*, 3629.
12. Golub, M. A.; Wydeven, T.; Johnson, A. L. *Langmuir* **1998**, *14*, 2217.
13. Golub, M. A.; Wydeven, T. *Polym. Prepr.* **1998**, *39* (2), 930; also this monograph.
14. Dilks, A.; Kay, E. *Macromolecules* **1981**, *14*, 855.
15. Norton, M. G.; Jiang, W.; Dickinson, J. T.; Hipps, K. W. *Appl. Surf. Sci.* **1996**, *96*, 617.
16. Sujimoto, I. *Macromolecules* **1991**, *24*, 1480.

Author Index

Subject Index

A

Acrylates, fluorinated monomers. *See* Acrylic-based polymers, fluorinated

Acrylic-based polymers, fluorinated
accelerated photoaging method, 132–133
change in infrared absorption at 1785 cm^{-1} in irradiated copolymers of butyl vinyl ether (BVE) with fluorinated and corresponding unfluorinated methacrylates, 139*f*
change of infrared absorption at 1785 cm^{-1} in irradiated 1H,1H,2H,2H-perfluorodecyl methacrylate/butyl acrylate (XFDM/BA) and 1H,1H,2H,2H-perfluorodecyl acrylate/butyl methacrylate (XFDA/BM) copolymers, 137*f*
changes in size exclusion chromatograms with degradation, 133, 136
comonomer alteration for copolymers of vinyl ethers with acrylic monomers, 136, 140
crosslinking of films of commercial Lumiflon LF200 upon irradiation in air, 140
effect of vinyl ether units on photodegradation, 136
experimental copolymers and structures, 131
experimental techniques, 132
fluorinated copolymers and unfluorinated analogues, 132*t*
gel content of selected fluorinated copolymers and unfluorinated analogues, 134*f*

irradiation in air of 2,2,2-trifluoroethyl methacrylate/BVE (TFEM/BVE) and ethyl methacrylate/BVE (EM/BVE), 136
modifying fluorine content and distribution along macromolecular structure, 130
normalized SEC analyses of irradiated copolymers of BVE with fluorinated and corresponding unfluorinated methacrylates, 138*f*
normalized SEC analyses of irradiated methyl α-trifluoromethyl acrylate/2-ethylhexyl vinyl ether (MTFMA/EHVE) copolymer, 141*f*
normalized SEC analyses of irradiated XFDM/BA and XFDA/BM copolymers, 135*f*
numbers of chain scissions per molecule in irradiated TFEM/BVE, EM/BVE, and MTFMA/EHVE copolymers, 139*f*
partially fluorinated side chains slowing photodegradation process, 133
polymer synthesis method, 132
production of unwanted, crosslinked, insoluble polymer as performance indicator, 133
photoinduced reactivity of vinyl ether units, 140
requirements for protective coating for stone, 130
selection of copolymers for fluorination, 129–130
sensitivity of vinyl ether units in

224

abbreviations, compositions, and
suppliers, 6*t*
market issues, 7
Perfluoroalkyl-grafted polysiloxanes,
side chain orientation reducing surface
tension, 5
Perfluorocarbons
amalgamated copper and mercury(II)
fluoride yielding, 2
atom bomb developments, 2
blooming effect, 4
blooming of mixture of hydrocarbons
and, 8
new refrigerants, 1
structural differences from
hydrocarbons, 2–3
surface tensions for, and analogous
hydrocarbons, 4*t*
1H,1H,2H,2H-Perfluorodecyl acrylate
(XFDA). *See* Acrylic-based
polymers, fluorinated
1H,1H,2H,2H-Perfluorodecyl
methacrylate (XFDM). *See* Acrylic-
based polymers, fluorinated
[3-
(Perfluorododecyl)propyloxy]triethox
ysilane (FDOPTES)
high resolution low pass filtered
atomic force microscopy (AFM)
images for FDOPTES monolayer,
37*f*
surface pressure-area isotherm and
electron diffraction pattern, 35*f*
See also Fluoroalkylsilane and mixed
monolayers
[2-
(Perfluorooctyl)ethyl]trichlorosilane
(FOETS)
surface pressure-area and electron
diffraction pattern, 35*f*
See also Fluoroalkylsilane and mixed
monolayers
Perfluorooctylethyl alcohol. *See* Fatty
acid esters
Perfluoropolyether polymers

chemical structure of lubricants, 85*f*
preferred lubricant material, 83, 86
See also Fluorocarbon polymer
mobility on disk surfaces
Phase behavior,
oligo(hexafluoropropene oxide)
substituted polymethacrylates
(PMAs), 73
Phase separation, preparing structures
of controlled size and morphology,
96–97
Phosphocholine
single-chain fluorinated as
surfactants, 49
See also Colloids with
fluorocarbon/hydrocarbon
diblocks
Photooxidation, sensitivity of vinyl
ether units in polymers, 136
Photostable coatings. *See* Acrylic-
based polymers, fluorinated
Piezoelectric properties, partially
fluorinated polymers, 3
Plasma deposited hexafluoropropylene
(HFP) films
adhesion force experiments by
pulsed force mode (PFM), 194
adhesion force measurements, 198–
199
adhesion forces by pulsed force
mode (PFM) as function of applied
reactor power, 197*f*, 198
advances in scanning force
microscopy (SFM), 188
advantages of radio frequency
plasma enhanced chemical vapor
deposition (RF–PECVD) for
biomaterial applications, 188
analysis of HFP polymer deposited
as function of reactor power, 197*f*
calculated crosslinking and
branching degrees, 197*f*
calculating degree of crosslinking,
195–196
cleaning and photolithographic

238

X